D1667573

WHOM THE GODS LOVE

CLASSICS
IN MATHEMATICS EDUCATION

Volume 1: *The Pythagorean Proposition*
by Elisha Scott Loomis

Volume 2: *Number Stories of Long Ago*
by David Eugene Smith

Volume 3: *The Trisection Problem*
by Robert C. Yates

Volume 4: *Curves and Their Properties*
by Robert C. Yates

Volume 5: *A Rhythmic Approach to Mathematics*
by Edith L. Somervell

Volume 6: *How to Draw a Straight Line: A Lecture on Linkages*
by A. B. Kempe

Volume 7: *Whom the Gods Love: The Story of Evariste Galois*
by Leopold Infeld

WHOM THE GODS LOVE:
The Story of Evariste Galois

Leopold Infeld

THE NATIONAL COUNCIL OF TEACHERS OF MATHEMATICS

1906 Association Drive, Reston, Virginia 22091

Copyright © 1948 by Leopold Infeld
Copyright © renewed 1975 by Maria Helena Infeld
All rights reserved

Library of Congress Cataloging in Publication Data

Infeld, Leopold, 1898–1968.
Whom the gods love.

(Classics in mathematics education ; v. 7)
Reprint, with new introductory material, of the ed. published by Whittlesey House, New York.
Biography: p.
1. Galois, Évariste, 1811–1832. 2. Mathematicians—France—Biography. I. Title.
QA29.G2515 1978 510′.92′4 [B] 78-3709
ISBN 0-87353-125-6

Printed in the United States of America

1978

FOREWORD

Some mathematical works of considerable vintage have a timeless quality about them. Like classics in any field, they still bring joy and guidance to the reader. Books of this kind, when they concern fundamental principles and properties of school mathematics and are no longer readily available, are being sought by the National Council of Teachers of Mathematics, which began publishing a series of such classics in 1968. The present title is the seventh volume in the series.

Whom the Gods Love: The Story of Evariste Galois was originally published in New York and London by Whittlesey House, a division of McGraw-Hill Book Company in 1948. The present reprint edition has been produced by photo-offset from a copy of the original publication. Except for providing new front matter, including a biographical sketch of the author and this Foreword by way of explanation, no attempt has been made to modernize the book in any way. To do so would surely detract from, rather than add to, its value.

ABOUT THE AUTHOR

Leopold Infeld was born in Kraków, Poland, then under Austrian rule, on 20 August 1898. Even though he was fascinated by physics at an early age, he reluctantly went to a commercial secondary school at the insistence of his father, a merchant. But he studied by himself to prepare for the *matura* examination, which was essential for anyone wishing to attend university. In this way he managed to graduate from the Jagiellonski University in Kraków.

Leopold Infeld's way to a career in physics was not easy. Even after he graduated from university, he could not find an academic post in Poland. He then went to Berlin and applied to take graduate courses at the university there. Here again he met with refusal (he believed it was because he was from Poland), but Albert Einstein arranged for him to attend lectures as a special student.

Back in Poland, he completed requirements for his Ph.D. degree at the Jagiellonski University, but again he could find no academic post even though at the time he held the only Ph.D. in theoretical physics granted in independent Poland after World War I. For eight years he taught at Jewish secondary schools. But he did research on his own and became a member of the Polish Physical Society. Moreover, his first popular science book, *New Pathways of Science,* appeared in Polish and later in English.

From 1929 to 1936 he was associated with the university in Lwów, where he became a docent. In 1933 he received a Rockefeller Fellowship and went to Cambridge, England, where he worked with Max Born. Born was known for his research on the particle and wave aspects of matter. Together they worked on a field theory that spanned modern quantum mechanics and electromechanic wave experiments of the nineteenth century.

After his return to Poland, Infeld again failed to become a professor; with fascism rising in Europe, he went in 1938 to the United States. There he received a fellowship at the Institute for Advanced Study in Princeton, again with the help of Einstein, with whom he worked on the relativity theory. This collaboration lasted for many years—indeed, until Einstein's death.

In 1938 Infeld went to Toronto, Canada, and was a professor at the University of Toronto from 1939 to 1950. When World War II broke out, he felt keenly the fate of his native land and his family (later he learned that his two sisters and other members of his family were killed by the Nazis). He did war work in Canada on radar but never on atomic energy, which was not his field. In 1949 he visited postwar Poland and was deeply moved by the horrible destruction of the war. He planned to take a leave of absence in 1950–51 and lecture at the University of Warsaw as a contribution to the intellectual life which the Nazis had tried to eliminate altogether. However, an attack was launched in Canada falsely stating that he would take the secret of the atomic bomb with him (although he had never worked on it; his own work on radar had been declassified). The result of the smear campaign was the threat of dismissal from the University of Toronto and the demand when he was in Warsaw that he return his passport to the Canadian Embassy. Sickened by the attacks on him, he did so.

Leopold Infeld developed in Warsaw an excellent institute ot theoretical physics, whose professors and graduates are highly regarded and frequently invited abroad to many countries.

After World War II, he felt that everything possible should be done to preserve peace and that this could best be accomplished by reducing the tension between the two world "camps." To this end he lectured on the need for peace while still in Canada and later in Poland. He was one of the signers of the Einstein-Russell Appeal as well as a founder of the Pugwash movement that resulted from it, whereby for the first time physicists from the United States and the Soviet Union, together with those from other countries, began a dialogue that was credited with helping to achieve the adoption of the Nuclear Test Ban Treaty.

Infeld wrote some hundred scientific papers, three popular science books, two autobiographical books, and the present volume. These include the best-selling *The Evolution of Physics* with Albert Einstein (Simon and Schuster 1938), *Albert Einstein and His Influence on Our World* (Scribners 1950), *Quest, the Evolution of a Physicist* (Doubleday Doran 1941), and *Why I Left Canada: Further Reflections by Leopold Infeld* (McGill-Queens University Press 1978). The last deals with his troubles in Canada, his later work with Einstein, and his ideas and experiences in other fields.

Leopold Infeld's son by his first marriage died during World War II in what is now Israel. His children by his last wife, an American college lecturer on mathematics, are a son, who is himself a theoretical physicist, and a daughter.

HELEN INFELD

Warsaw, Poland

Helen Infeld, Leopold Infeld's widow, is the daughter of the late William S. Schlauch. Schlauch, who was an active member of the National Council of Teachers of Mathematics, was elected honorary president of the Council in 1948. He taught mathematics in the New York City schools and at New York University.

Portrait of Evariste Galois when he was sixteen.

He dies young whom the gods love.

—MENANDER.

Evariste Galois was scarcely twenty-three or twenty-four years of age at that time; he was one of the fiercest Republicans.

—ALEXANDRE DUMAS (Père).

In France, about 1830, a new star of unimaginable brightness appeared in the heavens of pure mathematics . . . Evariste Galois.

—FELIX KLEIN.

TO MY READERS

It was in a small but famous American college town, soon after the fall of France; I sat with my friends, and we were trying to lighten our gloom by controlled drinking and by repeating in semantic variations the printed news and Churchillian slogans. What we said was in effect (though more modestly stated) that "Liberty cannot die in the country of its birth" or that "France was betrayed" or (though it would be most embarrassing if anyone were to put it this way) that "France, like the Phoenix, will rise from its ashes." Then we talked about the French scientists and their fate. I mentioned the name of Galois. One of my friends, a writer, asked who Galois was. I told him the story of Galois' life. He said: "This is an amazing story. You must write it down. Write a book about him." I replied that there was a war on and that I was busy. But he had an answer to that one. "If you are busy, you need relaxation. The only pleasure in writing is to do it for relaxation." I argued that there are only a very few sources and that much remains unexplained about Galois. My friend became even more enthusiastic: "This is fine. No professor living on footnotes will tell you that you are wrong. You can invent to your heart's delight."

Later I thought about all this while walking through our University Library and looking for week-end reading. I checked in the catalogue under Galois and found listed there Dupuy's essay, quoted always and everywhere when Galois' life is mentioned, then Bertrand's essay, written six years later, containing some new material, which, strangely enough, I never saw used or quoted anywhere; then I found in our library a two-volume book on Louis-le-Grand, the school to which Galois went. I took these books home with volumes of Louis Blanc's "L'Histoire de Dix Ans" and Dumas' "Mémoires."

During the next few week ends something happened to me which is almost impossible to explain to anyone to whom it has not happened, but something that seems natural to everyone to whom it has happened. I fell in love with France of the nineteenth century. During the years of the war the thought of France and Galois was for me and my wife a needed escape in time of fears, doubts, adversities. I spent all my free time in learning about Galois' life and his epoch. Indeed, in the story of Galois there are two central figures, both equally important: Galois and the people of France.

After I read all the sources available in the rich libraries of this continent, I learned from Professor Synge (who learned it from Professor Courant) that Mr. William Marshall Bullitt of Louisville, Kentucky, has for years collected photographic reprints of all documents relating to Galois and has a collection of everything that has been written about Galois. Mr. Bullitt graciously put his collection and his time at my disposal. This collection even contains items that either he or his research staff had discovered, items that were unknown and unpublished before. Strangely enough, anyone wishing to write about Galois in Paris would do well to journey to Louisville, Kentucky. True, the additional documents do not add anything essential to Galois' story, yet it is good to know that I saw all the factual material there is. Of course, it is possible that some undiscovered or unpublished memoirs of that time may throw additional light upon Galois' life, but this seems to me very doubtful.

Yet all these known sources explain only fragments of Galois' life. They are like short segments which can be connected into different life lines. The rest must be done by thought, imagination and guesses.

A reader of a biography would like to know beforehand whether the story presented to him is true (whatever such questions mean). Some regard fictional biography as a bastardly form of biography that should be prohibited by law. But the word "fictional" is used in at least two different meanings. A biography is fictional if the author stands above facts, if he uses his prerogative to change their order or deform them knowingly for an artistic effect. In this sense, my story is not fictional, for I did not take the liberty of changing well-established facts relating either to history or to Galois' life.

But there is also another use of the word "fictional." A biography is called fictional if the author connects known events with unknown events that he has invented so as to give a fuller and consistent (according to the author) picture of his hero's life; if the characters talk and use words that history did not record. In this meaning of the word, this biography is fictional. Yet when the story reaches its drama in scenes that matter the most, then there is almost nothing in them but the record of history. I believe I am justified in saying that in all essential features the story told here is a true story. I shall give a short account of the relation between truth and fiction (in the sense of inventing when history does not speak), after the story itself is told.

This is perhaps as good a place as any to thank at least some of those who helped me: Mr. Bullitt, whose kindness I mentioned a while ago; Professors Coburn, Coxeter and Schlauch who read the manuscript and made helpful comments, my friend S. Chugerman and the librarians of my University who helped me in my research. Michel Gram, a young Polish soldier and a poet, condemned to death by his illness, died in a Toronto hospital. I visited him regularly and we became close friends. Our endless discussions about Galois helped me to clarify the story. He seemed to be more interested in Galois' fate than in his own dwindling life.

I do not thank my wife because it would be like thanking myself. This book is as much hers as mine.

LEOPOLD INFELD

CONTENTS

I

KINGS AND MATHEMATICIANS

1 : 1811

IN 1811 a long-desired son was born to the Emperor of France and a long-desired son was born to M. Nicolas Gabriel Galois at Bourg-la-Reine. The hundred and one cannon shots that told anxious Paris about the birth of the King of Rome were taken up and repeated throughout the length and breadth of the Empire. In the archives of Bourg-la-Reine a document tells us that M. Nicolas Gabriel Galois, thirty-six-year-old director of a boarding school, came at one o'clock on the twenty-sixth of October to the mayor of the town, and, showing him a day-old infant, assured the mayor that he, and his wife, Adélaïde-Marie Demante Galois, were its parents and that they wished the child to be called Evariste. The King of Rome grew up with the eyes of France, and of the world, upon him; Evariste Galois grew up with only the eyes of his family upon him.

In 1811 the golden surface of Napoleon's Empire glared dazzlingly over Europe. At the court, the ladies were covered with flowers, jewels, and waving plumes; officers of the imperial household, generals, marshals, counselors of state, foreign ministers, glittered with decorations and orders offered to them by the victor or by conquered kings. From his throne the Emperor surveyed with an eagle eye the plumed and bejeweled circle of his Corsican clan, the new aristocracy he had created, and the old aristocracy which the splendor of his court had brought back from exile.

The more the Empire shone from the outside, the more it rotted from the inside. Spain and Portugal were not beaten. Russia violated the continental blockade against England. Daily the Emperor read the reports of his spies, counterspies, and counter-counterspies. His marshals grew in riches, fat, and flabbiness. The lines of the Empire were extended to the breaking point. Old kingdoms were replaced by indolent kingdoms under his Corsican clan. The Empire grew overripe for disaster.

New forces would soon come into play. They were to shape the destiny of Evariste Galois, perhaps the greatest mathematical genius that ever lived.

What were these forces?

An old mathematical tradition flowered in France. Lagrange, Legendre, Laplace, and Monge created the mathematical atmosphere not only for France, but for the whole world. They were to influence Cauchy, Galois, and the generations of mathematicians to come. But this was not the only force and these were not the only men who influenced Galois' life. There would be new rulers in France—small, bigoted men would come to thwart and stifle Galois' genius.

There were Frenchmen all over Europe whose sorrows were the Empire's increasing glories. Their lawful King, brother of Louis XVI, was a victim of the same terror that had abolished nobility and driven the flower of France into alliances with foreign powers. For them the land of their birth and their possessions was now a hostile land under the yoke of the Corsican gangster. But some day the true sons of France, screened by the bayonets of foreign soldiers, would return to the old France of Henry IV and Louis XIV.

Louis XVIII held his court in Hartwell some fifty miles from London. He walked with difficulty, for his weak legs could not support his stout body. His manners were gracious; his eloquence cultivated; his remarkable memory was stuffed with Horace's verses and the remembrance of suffered wrongs and insults. Misfortune had never crushed him because a rigid royal armor stood between him and fate. He was greedy for courtiers. "Yet he was a king always and everywhere, as God is always and everywhere." This infirm man emanated a pride, dignity, and majesty that gave him power over human souls.

Later, even Bonaparte's generals confessed that they were more intimidated by the corpulent, grotesque Louis XVIII than by the Corsican who had led them to victories and disasters.

Comte d'Artois, the younger brother of Louis XVIII and the future Charles X, was a silly, scheming, elderly man, who spent his days and his money in England. This most accomplished gentleman among the exiled French nobility, famous in his youth for his elegance and love-making, was indeed a thoughtless fool, impatient of study, unrestrained in the indulgence of his desires and passions. He felt superior to others because the royal blood of the Bourbons flowed in his veins. In London, he confessed to an English friend, "I would rather be a cabdriver than King of England. I would not accept the crown of France at the price of a charter or any kind of constitution."

Where lay Napoleon's superiority over the two Bourbons—Louis and Charles? Napoleon's great superiority lay in understanding the obvious, so often hidden from twisted minds born to the purple and taught from childhood that nations revolve around the axis of their own lives.

Napoleon was mean, arrogant, dishonest with others and still more dishonest with himself, incapable of self-criticism, love, loyalty, or devotion. Yet he was the first among rulers to understand the simple truth that science is not a luxury that makes the Empire shine before the outside world; he knew that science wins wars too! He wanted the Polytechnical School, pride of his Empire, to grow and flourish, not only in time of peace but also in time of war, because one "must not kill the hen that lays the golden egg." He made the princes of the kingdom of mathematics dukes of the Empire, and friends of the Emperor.

Napoleon said, "The advancement and perfection of mathematics is intimately connected with the prosperity of the state." The history of mathematics and the history of mathematicians is only a sector of any history. Kings and mathematicians do not live in isolation. Mathematicians have been made and broken by kings. The lessons that history teaches us are obvious. Yet how many rulers have known or know that "the advancement and perfection of mathematics is intimately connected with the prosperity of the state"?

If we look at the heritage of Euclid and Newton from the perspective of the early nineteenth century, we see this heritage in its greatest bloom and glory. Yet, as viewed from the perspective of our present day, we see that this heritage, like the Empire, was ripening for a new order, for new ideas that would change our picture of the external world. Lagrange and Laplace! These two names symbolize better than any others the perfection and the end of the mechanistic philosophy that tried to disclose the past and the future of our universe.

The year 1811 saw old Lagrange nearing his grave. The aged man was serene, lonely, understanding and skeptical. He listened with a half-friendly, half-ironical smile to Napoleon's judgments on mathematics, history, and the state. He knew that the rulers of the world seldom have any doubts; that they succeed only because their ignorance is mixed with still greater arrogance. His own life had taught him that, unlike a king, a mathematician succeeds only if he has doubts, if he tries humbly and incessantly to diminish the immense extent of the unknown.

Lagrange's celebrated work *Mécanique analytique* crowns Newton's classical mechanics, builds it up formally to a structure as beautiful and rigorous as geometry. Lagrange said about Newton that he was not only the greatest but the most fortunate among the savants, because the science of our world can be created only once; and Newton created it!

In this same year, Laplace, the son of a peasant, was sixty-two years old and had become the Comte Pierre Simon de Laplace. The great Revolution had given him distinctions and honors; the Consulate had made him Minister of the Interior; the Empire had made him a count; the Restoration was to make him a marquis. Laplace, the little man and great snob, was an illustrious savant, to which fact his *Mécanique céleste* will remain an everlasting testimony.

Napoleon created an empire; Laplace created a consistent mechanical picture of the entire universe. In it a gigantic machinery operates eternally, its motion forever prescribed. Laplace's universe is deterministic. Indeed, if we know the state of the universe now, at this moment, that is, if we know the positions and velocities of all particles, all planets, all stars, if besides that we know the laws of nature, then we have in our hands all the knowledge by which to determine the future and the past of our universe. What *did* happen and what *will* happen are determined by what *is* now and what *are* the governing laws. If we know them, then the future and the past become an open book. There is nothing that must remain hidden forever from the human mind. The aim of science lies clearly mapped before us: to learn more and more about the initial conditions; to know better and better the laws of nature; to penetrate deeper and deeper into the mathematical formalism. These are the keys that will open the past and the future of our universe.

Arrogance governed the empire of science. The proud knowledge that the world is ruled by deterministic laws was to be destroyed one hundred years later by the advent of Quantum Theory.

But in 1811, both the empire and the deterministic view seemed at the peak of their power.

Even the Emperor had turned the pages of Laplace's *Mécanique céleste*. He was fascinated especially by its third volume, that is to say, by its dedication to "Bonaparte, the pacificator of Europe, the hero to whom France owes its prosperity, its greatness, and the most shining epoch of its glory."

In vain did Napoleon look for a similar dedication in the fourth volume. Impatiently he turned its pages full of formulas and calculations. He closed the book, sure that he had seen enough of it to expound his own views on the universe whenever it would please him to do so. Such an occasion came soon at a ball in the Tuileries, where he discovered a circle of savants clustered about Laplace, who was displaying all the splendors of the Grand Cross of the Legion of Honor and of the Order of the Reunion.

"Yes, Comte de Laplace. I just glanced again through your volumes on the universe. There is something that is missing in your important work."

"Sire?"

"You forgot to mention the Maker of the universe."

The count bowed and a sly smile passed over his face.

"Sire. I did not need this hypothesis."

The Emperor looked proudly at the clever savant. There was no pleasure in tormenting men who submitted too easily. He turned an examining eye toward Laplace's neighbor, an old man with hollow cheeks and a big nose.

"And you, M. Lagrange, what do you say to that?"

The tired old eyes lighted. "Sire. This is a good hypothesis. It explains so many things."

A powerful voice cut in. "Laplace's universe is as precise and efficient as a good watch. If one discusses watches, one does not need to discuss the watchmakers, especially as we know nothing about them."

Napoleon turned to the man who spoke and stared at his friend as though he would like to bore two holes through the broad ugly face with its disfigured flat nose. But the small eyes in the big face looked straight back.

"Ah, M. Monge! I should have known that you would not keep quiet when we are on the subject of religion. So, M. Monge, you think that the watchmaker does not need to be mentioned. Unfortunately, I am sure that many of your students in my Polytechnical School would agree with their beloved teacher."

He turned his eyes away from the inventor of descriptive geometry and shot words at all of them. "I, as the head of the great Empire, wish that you gentlemen, who enjoy my esteem and friendship, would put away once and forever your atheistic past which not all of you seem to have forgotten. The time of revolution is past."

Keeping one hand under his white vest, the other behind his back, he gently scratched the two sides of his body and instructed his audience, "I have restored the *priests,* though not the clergy. I wish them to teach the word of God so that it will not be forgotten. Please remember, gentlemen, that a moderate religion has and will have a place in my Empire."

He waited neither for an answer nor a reaction, but turned abruptly away to bully the rest of his guests and to receive their flattery.

Lagrange, Laplace, Monge. They were old men. Those who would soon build new foundations of mathematics were still unknown in France. They were: Augustin Louis Cauchy and Evariste Galois.

On the Atlantic coast, a wall was being built against a possible British landing, and port installations for the invasion of England, scheduled to follow the defeat of the Russian Czar. One of the small wheels in the machinery of defense was Augustin Louis Cauchy, who in the years to come would break the spell of Newton's heritage and establish links to modern mathematics. In 1811, the twenty-two-year-old Cauchy worked from early morning until evening on fortifications at Cherbourg in the service of the Emperor, whom he would soon learn to hate and despise.

His nights were free. Almost every evening, Cauchy wrote to his mother letters full of love and devotion. After finishing his letter, he would turn to the books lying on his tiny table: Laplace's *Mécanique céleste,* Lagrange's *Traité des fonctions analytiques,* and Thomas à Kempis' *Imitatio Christi.* He knew he would become a great mathematician. Was he not the best student in that subject at the Polytechnical School, and did not M. Lagrange predict that some day he would outrank the greatest living mathematicians? He would not disappoint M. Lagrange. He would think everything anew, from the beginning; he would make methods of proofs and reasoning simple, clear, and convincing.

In Cherbourg, Cauchy wrote his first paper, on the subject of stone bridges. It was lost in the pocket of the secretary of the Academy and it was lost to posterity, since Cauchy did not keep a copy. Thus, as a young man, Cauchy learned that it is consistent with the laws of nature for scientific papers to vanish into thin air and never be seen again. But Cauchy had trust in God and in himself. From stone bridges he turned to mathematics, and in a few years he was to become the greatest mathematician of France, and in all the world of living mathematicians, second only to Gauss.

We shall meet Cauchy later. We shall see how his life was bound up with those of the Bourbon kings, Louis XVIII and Charles X.

In 1811, and thereafter, history recorded fully the birth of the King of Rome, his transition to the Duc de Reichstadt, the hatreds, fears, and intrigues of kings, and his journey toward death.

The story of Evariste Galois, his hatred of a king, his journey toward death and fame, is recorded only in fragments, disconnected, suppressed, and contradictory.

2 : 1812–1823

In 1812, Napoleon's famous twenty-ninth bulletin told the Parisians for the first time that their great army had been annihilated by the Russian cold, though not by the Russian soldiers. The French thus learned with stupefaction and anger that their Emperor was not invincible.

A flood of Russian, Prussian, and Austrian soldiers moved over Europe, toward the Elbe, toward the Rhine, toward the Seine, drowning on its way Napoleon's tyrannies and restoring the tyrannies of the old kings.

In 1813, Lagrange died at the age of seventy-seven. It is difficult to imagine that during these gloomy days Napoleon, who had seen death thousands of times, would mourn the passing of a gentle old man. Yet, if we are to believe the Duchesse d'Abrantes, the Emperor was "deeply affected." He was supposed to have said, "I cannot master my grief. I cannot account for the melancholy effect produced upon me by the death of Lagrange. There seems to be a sort of presentiment in my affliction."

Lagrange's own last words, spoken to Monge, are recorded by history:

"Death is not to be dreaded. When it comes painlessly, it is not unpleasant. In a few moments my body will cease to live and death will be everywhere. Yes, I wish to die and I find it pleasant. I have lived my life. I have achieved some fame as a mathematician. I never hated anyone. I did no wrong, and it will be good to die."

Nations should build momuments to men who can justly say such words on the threshold of that experience which is common to all, but imagined by none.

The Empire survived Lagrange's death by only a year. Twice did Napoleon try during that time to halt the onrushing flood, and twice he failed. France became sick and tired of blood and broken promises, of a spectacle that had been exciting but had become boring. The Parisians were almost relieved when Napoleon left them alone, and when the Russian, Austrian, and Prussian armies entered their city. Now shops were open and crowded by foreign officers; Russians drank punch in the cafés; Cossacks built huts in the Champs Élysées. General Blücher wore seventeen medals upon his breast, and Parisians were impressed by so much evidence of military genius. The Parisians stared at the new scenery and accepted it with little remorse or sorrow.

France was made safe for the Bourbons. Napoleon abdicated, and Louis XVIII and the entire Bourbon family returned to France. The long-forgotten rulers were greeted with noisy assurances of loyalty. White, the color of the Bourbon flag, became the color of Paris; Bourbon lilies adorned French women, and even in the suburbs people hung out dirty sheets from the windows.

When, on the third of May, 1814, Louis XVIII entered Paris, the Quai des Orfèvres was lined from the Pont Neuf to Notre Dame with a regiment of the old guard to screen the foreign soldiers from the King's sight. These grenadiers, who reeked of fire and powder, to whom Napoleon was a godlike hero, were forced to salute a king whose infirmities were those of age and birth and not of war. Some of them contrived, by wrinkling their foreheads, to shake their large busbies over their eyes so that they would not see; others, through their mustaches, bared their teeth like tigers. They presented their weapons with a movement of fury that put fear into the hearts of men and women lined behind the grenadiers to wave their white handkerchiefs and to shout: *Vive le roi! Vive notre père!*

When Louis XVIII arrived at the Tuileries, he exclaimed, smacking his lips at the unexpected richness and splendor, "He was a good tenant, this Napoleon."

Soon after the King was enthroned, the flood of foreign soldiers receded, and Louis XVIII convoked the chambers to grant the paternal charter to his children. He was dressed in a uniform that was his own artistic invention, designed to cover his infirmities and add dignity to his overfed frame. The coat of blue cloth was something between a frock coat and a court dress. The two gold epaulets made of it something between a court dress and a marshal's uniform. His satin breeches ended in red velvet shoes that came over the knees. His gout did not allow him to wear leather, and the King always made a great point of his shoes. He believed they made him look as though he could at any moment mount a horse and repel the invader. If the boots were not convincing enough, there was also a sword that hung at his side. His face was powdered and his hair carefully curled over his ears. In this dress and against a background of Roman statues, the King read his speech in a sonorous, smooth voice as his double chin quivered and his blue eyes wandered leisurely from the document to the audience.

He said that he was very well satisfied with himself; he congratulated himself for being the dispenser of the benefits that Divine Providence had designed to grant to his people. He congratulated himself on the peace treaties signed with European powers, on the glories of France's army, and on the happy future of France that his old eyes foresaw. There were no longer any clouds; hosanna for the new ruler, whose only ambition was to carry out the wishes of his brother, Louis XVI, as expressed in the immortal testament before his head was cut off! These were, he assured his listeners, his paternal intentions.

When the King finished, M. Dambray, the chancellor, read a commentary to the charter, the new constitution of France.

"The breath of God has overthrown the formidable colossus of power that was a burden to the whole of Europe, but under the ruins of that gigantic structure France has discovered the immovable foundations of her ancient monarchy. The King, being in the full possession of his hereditary rights over this noble kingdom, will only exercise the authority he holds from God and his father, within the limits that he himself has set."

Thus, it was not the nation that extracted the charter from the King; it was the King who gave it in fatherly love to his people.

Against the same background of Roman statues, the same actors enacted a very different scene a year later. It was the day when the new throne of the Bourbons trembled to the rhythm of Napoleon's marching soldiers. The marshals, generals, and the old guard broke their oath to their king, unable to resist the old charmer who, after landing on French soil, again engulfed his country in blood. The unhappy Louis appeared at the Palais Bourbon surrounded by dignitaries of the court and all the princes of his house. The shrewd old man knew that the charter he hated, but had been forced to sign, was his only trump card. Thus, said the King, Napoleon came back to take away the charter and liberties from the French people—"the charter that is dear to the heart of every Frenchman and that I swear, here and now, to maintain. Let us rally around it! Let us make it our sacred standard."

The chamber resounded with cries:

"Long live the King! We will die for the King! The King forever!"

Then Comte d'Artois walked toward his brother the King, turned toward the peers and deputies.

"We swear to live and to die true to the King and the charter, which assures the happiness of our fellow countrymen."

The two brothers fell into each other's arms as the spectators wept and shouted, *"Vive la charte!"*

Fifteen years later, we shall hear the same cry resounding throughout France, and we shall see people die in Paris because a king broke his oath.

Louis talked about dying on the battlefield, but when Napoleon swept through France, the King and his brother fled to Belgium. They came back once more to a Paris full of foreign soldiers after Napoleon had lost his last battle at Waterloo and with it his hundred-day-old crown and freedom.

Now the Bourbons saw how wobbly was their throne, and how little the French children cared for their kingly father. How could they attach the Bourbon throne to the shifting soil of France? The King and his courtiers had the answer: by prison walls and blood. The white terror of the Restoration began.

Even mathematicians were not spared.

Gaspard Monge, the son of a peddler and knife grinder, had embraced the cause of the Revolution and was the savior of Republican France, when, together with Berthollet, he organized the manufacture of gunpowder. Monge loved the Republic, but he loved Caesar more. When he became Comte de Péluse, he forgot how loudly he had shouted for the abolition of nobility before Caesar became Ceasar.

Gaspard Monge was a great mathematician and a great teacher. He was the inventor of descriptive geometry, and he was the man who organized the Polytechnical School and laid the foundation of its great tradition. He is the father of modern teaching of mathematics throughout the civilized world. The modern textbook takes its origin from the published lectures at this famous school, the first great school of mathematics on our globe; a school that educated future officers, savants, statesmen, rebels; a school admired by scientists and feared by kings.

But Monge had committed a crime: during the hundred days he had remained faithful to Napoleon. In 1816 Monge was seventy years old. He had retired from the Polytechnical School because his hands were partially paralyzed; then, while reading the tragic twenty-ninth bulletin, he suffered a stroke and never fully recovered. It would perhaps have been merciful to kill the old man. This the King did not do. But he expelled Monge from the Academy. Under this blow, Monge, driven by fear, went into hiding where, sick and lonely, he died two years later praising Caesar and cursing the Bourbons. But the revenge of kings goes beyond death. The pupils of the Polytechnical School, to whom Monge's name had become a legend, were forbidden to attend the funeral of their great teacher.

Monge had been expelled from the Academy and his chair vacated. Was there in France a mathematician with so little decency as to accept Monge's chair?

In 1811 we saw Cauchy in Cherbourg. Five years later he, by then the greatest mathematician in France, was offered Monge's chair at the Academy. He accepted it readily; in the same year he became a professor at the Polytechnical School. It was a good year for Cauchy. The King was kind to the great mathematician, and Cauchy was to show posterity some years later that the Bourbons, too, can be loved and admired by an illustrious savant.

If it is the duty of kings to punish loyalty to fallen masters, then it is their privilege to reward treachery. In the same year that Cauchy became a member of the Academy, the great mathematician and astronomer, Laplace, was entrusted by the King with the task of reorganizing the Polytechnical School so as to bend the unruly students to the will of the crown. Laplace deserved the King's confidence. He offered his services to the Bourbons; as a senator he signed the decree banishing Napoleon; he even, at his own expense, changed the dedication in his unsold volumes on probability from "Napoleon the Great" to "Louis XVIII." But did Laplace succeed in spreading love for the Bourbons among the young students of the Polytechnical School? We shall see—fourteen years later—that he failed miserably. The waves of history, whether ascending or descending, raised Laplace, the peasant's son, until they landed him as Marquis de Laplace in the Chamber of Peers. His old eyes had seen the *ancien régime*, the Republic, the red terror, the Consulate, the glory and wretchedness of the Empire, the white terror, the appearance of five volumes of his *Mécanique céleste,* and the return to power of the aristocracy, the "ultras" under the two last Bourbon kings.

The ultras, that is the ultra-aristocratic supporters of the King and the ultra-zealous defenders of the Church, were hated by the people and by the still powerful bourgeoisie. Even the King hated them at the beginning of his reign. But, as the impotent King became older and weaker, the brother, "Monsieur," and the ultras around him became more unscrupulous and more successful. It is possible to set almost the exact day on which the ultras took power and the single event that threw it into their hands.

The King had no children. Monsieur, his brother, the future Charles X, had two sons. One of them, the Duc de Berry, was a rough, uneducated playboy who intrigued with his father against the King and in fits of temper tore epaulets from the uniforms of his officers. The other, the Duc d'Angoulême, was weak, ugly, shy, but not devoid of all noble instincts. Both of them were married; neither of them had legitimate children. If they should die, the tree of the Bourbons would be uprooted from the soil of France.

On February 13, 1820, the opera on rue de Richelieu played *Le Carnaval de Vénise* and *Les Noces de Gamache* with new dancers. The Duc and Duchesse de Berry were present, but the duchess did not feel well and wished to leave before the end of the performance. As the duke was assisting Her Royal Highness into the carriage, a man collided violently with him and ran away without apology.

"What a ruffian."

Then, with a queer mixture of astonishment and terror, the duke cried out, "I am stabbed!"

He was helped to the antechamber of his box and then, as the wound looked serious, to the manager's office. Soon came the duchess, Monsieur, the courtiers, ministers, doctors, the captured murderer with hands and legs bound, the police escort, the bishop.

The duchess cried hysterically, mixing French with her native Italian. Through calm intervals between her shrieks one could hear music from the opera orchestra, weak singing voices, and the applause of the audience. These irregular sounds were then covered by the bishop's monotonous recital of Latin prayers.

In the corner of the room, Monsieur and the Prime Minister argued whether to call the King. The father of the dying man said that court etiquette did not allow the King to come into a manager's office. The groaning duke turned toward his father.

"I want to see the King." Then to the screaming duchess, "Be calm, my dear. Think about our child."

At the last words, faces lifted and eyes glanced.

The bishop forgave the duke his sins.

"I want to see the King. I have two daughters. I want to see my two daughters. You don't know. Send to their mother, Mme. Brown."

He stammered the name and address, known to everyone anyhow. Messengers were sent and reappeared with two frightened girls, squeezing them through the crowded room. The duke smiled at them and did not protest when they were soon taken away. By now he was sinking fast and only repeated mechanically, "I want to see the King."

At five o'clock in the morning, the King's wheel chair with the King in it was pushed with great difficulty up the narrow stairs leading to the manager's office. The porters groaned and sighed under the heavy load. When the King was safely landed near his nephew, the duke became suddenly conscious and his words were clear.

"Forgive me, my uncle. I beg you to forgive me."

The tired King breathed with difficulty.

"There is no hurry. We will talk about it later, my son."

The last spark of life and terror lightened the duke's eyes.

"The King does not forgive me—my last moments—not softened by forgiveness."

These were the last words of the Duc de Berry. The doctor asked for a mirror. Louis handed him a snuffbox and the doctor held it to the nostrils and lips of the duke.

The doctor whispered, "It is all over."

The King said to the doctor, "Help me, my son. I have one last service to render."

The crippled old man leaned on the doctor's arm, closed the eyes of the corpse that had been the dashing Duc de Berry, and all present fell on their knees.

The wretched man who assassinated the duke achieved nothing. The entire opposition was now abused as the accomplices of murder. The ultras shed tears and threw slander until the power was theirs.

Seven months later, a son was born to the Duchesse de Berry. Smart and fashionable Paris repeated joyfully, "He is born, the child of a miracle, the heir of martyr's blood."

The memory of this child will disturb Paris ten years later. But he will never become a king of France.

In May of 1821, news vendors cried on the streets of Paris:

"Napoleon's death and his last words to General Bertrand!"

There was little interest. Since 1815, the old Emperor had been forgotten by the French people. Yet in a few years he would live again. A new Napoleon was in the making: a Napoleon dressed in his plain gray coat, chatting with his soldiers around the bivouac fire; a Napoleon loving peace and his French people, but forced to fight because of perfidious Albion's intrigue; a Napoleon assassinated by the English oligarchy at St. Helena, whose last wish was "that my ashes may repose on the banks of the Seine, in the midst of the French people whom I have loved so well." Napoleon's legend was on the march!

One of the most striking aspects of the Restoration epoch is the influence of a society that officially had been banned from France: the Society of Jesus. Its dense and well-woven web covered the political and educational life of the country. The Jesuits, or their supporters, were among the clergy, among the deputies and peers, in the Faubourg St. Germain, among cabinet ministers and Monsieur's courtiers. These lay sympathizers, "the Jesuits of the short robe," were either everywhere, or believed to be everywhere, by those who disliked the ultras. We are told that spies working for the Jesuits were among valets, janitors, chambermaids, and in the police force; that under the guidance of this order, new laws were introduced into the chambers; that even the King in the last years of his reign became a tool in their hands. This was supposed to have happened not by chance, but by a carefully planned intrigue.

Until the assassination of the Duc de Berry, the King had always been under the influence of men favorites—always one at a time. The last one, Décazes, was Louis' Prime Minister and the man most hated by the ultras. The King was forced to dismiss his favorite when slander, curses, and accusations were thrown at him after the death of the Duc de Berry. And, for the first time in the King's life, it was a woman that consoled the heartbroken old man.

Mme. du Cayla was brought to the King's attention by Father Liautard, a member of the Society of Jesus. About the Father's social views we do not need to guess. They were put forward clearly enough in ink in an essay *The Throne and the Altar.* There we are persuaded by Father Liautard that the public press should be abolished; that there should be only one daily paper, sponsored by the King and edited by the Chief of Police. It should contain interesting and useful news, like the variation of temperature and the prices of wheat, coffee, sugar. This, Father Liautard assures us, would meet all reasonable requirements. At the same time, the poisonous books of Rousseau and Voltaire should be burned.

In 1821, Mme. du Cayla's influence upon the King was well established. Having been carefully tutored by Father Liautard, she knew how to amuse and interest the King. Willingly she accepted orders from Monsieur, the clergy, and the aristocracy of Faubourg St. Germain. The web spun tighter and tighter until the King became a tool in the hands of his brother.

Soon the web began to cover the schools of France so as to squeeze out of them the spirit of rebellion. The Polytechnical School came first. It had been reorganized and in it the atheist, Monge, replaced by the pious Cauchy. The Normal School also was the daughter of the Revolution, and its purpose was to prepare masters and professors for the Royal Colleges. It was closed in 1822 and, in this simple manner, a possible breeding place of Republican and Bonapartist ideas uprooted. Then the web began to be spun around the Royal Colleges. Among them the greatest and most important was the College of Louis-le-Grand. Its purpose was to educate polished gentlemen versed in Latin and Greek, and, above all, loyal subjects of the King and defenders of the Church. Did the school achieve its aim? Let us, in answer, look for the names of its three most illustrious students.

Through Louis-le-Grand had passed the "incorruptible" Robespierre, who offered to the guillotine the head of Louis Capet, the former Louis XVI.

Through Louis-le-Grand had passed Victor Hugo, who later fought the tyranny of Napoleon III, whom he christened for posterity "Napoleon the Little."

In 1823, Evariste Galois, having been carefully tutored by his mother, passed the examination for Louis-le-Grand and entered its fourth class. He, too, hated and fought a king of the French.

II

THE REBELLION AT LOUIS-LE-GRAND

1 : Sunday, January 25, 1824

A THIN MAN with tightly closed lips moved silently across the study toward M. Berthot's desk. With one hand M. Berthot indicated a wooden chair and with the other he nervously milked his graying red beard.

"It is good that you came, M. Lavoyer, very good indeed. I knew you were one of the few parents on whom I could count."

He took a pinch of snuff, pushed it into his big nostrils, and turned confidentially toward his guest.

"We have a pretty full picture now, and we know what to do, with the help of loyal parents. Unfortunately there are only a very few indeed on whom we can count. Yes, M. Lavoyer, you know how to deal with conspiracies. Please tell me your story."

M. Lavoyer looked at the floor and his voice was humble.

"There'll be a revolt on Tuesday, the day after tomorrow, at six o'clock in the evening."

M. Berthot leaned back in his black leather chair and sighed deeply.

"Yes, I know." Then, clenching his fists:

"The leaders! We must know the leaders. All of them."

He banged his clenched fists on the desk.

M. Lavoyer took a neat envelope from his pocket and placed it silently on the desk. With short thick fingers M. Berthot put a piece of paper from the envelope beside a long sheet lying on the desk. His eyes jumped quickly from one page to the other while he made notes and checks.

"Ha! I thought so. Yes, quite right. I'll show them. We shall see, we shall see." Then he turned to his guest. "We are getting somewhere now. Other lists together with yours give us about forty names. We have all the leaders now! And what names! The best students at Louis-le-Grand. Students to whom we have given prizes and for whom we have done most. They have been poisoned by outsiders, and they have carried this poison into the school. They would like to bring Napoleon back from his grave. But this is not the worst. Some of them would like to bring Robespierre from his grave!"

M. Lavoyer shook his head sympathetically backward and forward.

"Now let's go into detail. We told your son that he needed rest when we sent him home. How did you get all this information from him?"

The thin man froze.

"Sir! I don't want to discuss that. My son doesn't know that I'm here."

"Do not worry about your son, M. Lavoyer. He is a good boy, a good student, and he works hard. I can assure you that he will get a scholarship next year. He will receive the finest education in France without cost, without any cost at all. I can promise you this for as long as I am here."

He brooded, then shot out angrily, "What do they have against Louis-le-Grand? That is what I would like to know."

"Sir, it's a painful subject; I'd rather not say."

"But you must tell me. I must know. I insist!"

M. Berthot's bloodshot eyes were fixed upon the rigid face opposite him.

"They say that you, sir, will bring the Jesuits back and give the school to their order."

"Oh! Again that silly old tale." He spoke with bitterness and self-pity. "I, who cannot appoint a professor, or expel a single student without the minister's signature, I am supposed to be delivering the school to the Jesuits. Yes, I know. To these Bonapartists and Republicans, everyone who is loyal to our King is a Jesuit. They would like to frighten everyone with the Jesuits. What else do they say?"

The smooth, monotonous voice answered, "They say the food's bad and they don't like the cooking."

"Another old story. They complain about food. I wish I could lay my hands on these outsiders who are stirring up trouble at Louis-le-Grand. Bad food, indeed! Is there something else too?"

His intense voice had an overtone of fear.

"There is something else, but I don't like to say."

"Let us have everything. I know that this conversation is not very pleasant for either of us."

M. Berthot opened and closed the same button of his black, greasy coat. M. Lavoyer's eyes suddenly lit up as he recited, "They'd like you removed, sir, because, they say, your dress and manners insult the school."

A purplish red spread over M. Berthot's face, deepest at his neck where the red contrasted with the black of his wornout tie. He stopped unbuttoning his coat and tried to steady his trembling fingers by drumming on the desk.

"I'll show them; I'll show them!"

With disgust and shame he looked at the wax figure opposite him, which he now hated almost as much as his students.

"You have been very helpful, M. Lavoyer, very helpful indeed. Thank you very much."

All color had drained from M. Berthot's voice. He stood up, pushed the chair from his desk and stretched out his hand to M. Lavoyer. The thin man bowed and noiselessly closed the door behind him.

Louis-le-Grand was silent at nine-thirty that evening. A master stood outside each huge room where the students slept. With one ear pressed against the door, each tried to hear the subdued conspiring voices. This was the time to prove his loyalty, to supply information, to increase his salary from twelve hundred to fifteen hundred francs a year, and to lay the groundwork for a future career as a teacher or even a professor at Louis-le-Grand.

All the masters who were not on duty, all professors, subdirectors, and the proctor were assembled in the long conference room where the air smelled of snuff and tobacco. M. Berthot sat at the head of the table, whose green cover was stained by ink and wax. Around it the first ring was formed by about forty professors; the second, at a respectful distance, by about forty masters.

The director waved a bell sharply with his snuff-stained fingers and spoke in a thick, loud voice, spitting small drops of saliva upon the table and his neighbors. ("Don't come near the director without an umbrella" was a standing joke among the masters.)

"Gentlemen! This is a grave hour in the history of our school. There is a very grave danger that the events of the terrible days of 1819 may return to Louis-le-Grand. This we must prevent at all costs!

"You will hardly believe me, but I tell you, gentlemen, it is all true. Terrible things are planned, and you are fortunate, gentlemen, that I found them out just in time."

The eyes of professors and masters that were turned toward the director saw with astonishment that he now wore his best blue coat, a clean white shirt, and a broad black tie shining with newness.

The director pointed toward the second ring formed by the masters.

"The students plan to beat you, throw you out of the windows, and smash the furniture."

The director pointed toward the first ring formed by the professors.

"You too are to be thrown out. They plan to take possession of the whole school and then, they think, they can dictate the conditions of peace."

He glowed with the sense of power that could spread the waves of fear. Now it would be doubly pleasant to calm these swelling waves and to uncover the firm rock of authority.

"They made only one mistake. They forgot that there is a director in the school. I now know all the leaders of the rebellion. I know each one of them. I have all their names! There are forty of them, gentlemen!"

A weak, ugly smile spread over his face.

"I can assure you, gentlemen, that I know what to do with these rebels. You will learn about it in due time. With your help I shall save the college of our beloved King."

The director turned with a gesture of disgust and impatience toward his right-hand neighbor. He had never liked this M. de Guerle and he never would. The man seemed old, tired, and weak, but the calmness of his obstinacy was unbreakable. How had this relic of Napoleonic times survived at Louis-le-Grand? At a school that was once proud to belong to the Jesuit order? He should have known that a teacher admired by the students is bad and dangerous. He would get rid of the rebels; he would get rid of M. de Guerle.

"Our proctor, M. de Guerle, second only to me at Louis-le-Grand, has asked me for permission to make an announcement in his own name. I have agreed, but I wish to tell you as the director that I do not approve of what he will say."

M. de Guerle rose and began to speak quietly, almost in a whisper.

"This is my fifteenth year as proctor of Louis-le-Grand. I lived through the terrible days of 1819. I saw our school closed, dissolved; I saw hatred and mistrust spread. I shall never forget those terrible days. No! I don't think the students have the right to interfere with the school's administration. But neither do I believe in the use of force. What we are experiencing today comes from the use of force five years ago. Perhaps tomorrow we shall think we have succeeded. But some years later we'll find out that what we really did was to sow the seeds of future revolt."

He spoke calmly, undisturbed by the director's attempt to scratch a wax stain from the green table cover.

"The present cry of the students is 'away with the Jesuits.' I agree that their cry is unreasonable and I agree that, unfortunately, our students show little religious feeling. I would even go so far as to agree that someone may have put dangerous Republican ideas into their heads. But how did it happen that these outside influences could organize all the students? Does it not show that they are unhappy and dissatisfied here at Louis-le-Grand? Their unhappiness and dissatisfaction may spring from their imaginations, I agree. But then what do we accomplish by force and terror?"

He paused, and when he started to speak again his voice was even calmer.

"I am afraid that I am defending a lost cause. If so, then this is my last year at Louis-le-Grand. But I would like to suggest something. We have here the names of forty leaders. Why not assemble them tomorrow morning, to hear their demands and try to persuade them to be reasonable? We may be able to reach a compromise and thus save the school and ourselves from disgrace.

"Gentlemen, I know that my words will sound strange to you. But we cannot win this fight. We cannot win a fight against nine hundred students. We may seem to achieve victory. But the greater the victory seems to us, the greater will be our final defeat."

The director had been drumming on the table, and before M. de Guerle sat down, he began violently:

"If I understand you correctly, M. de Guerle, you want us to negotiate with the rebels, treat them as the equals of professors and masters. If they say they don't like M. Berthot, or they don't like these professors or those masters, you would tell them, 'Very well, boys, it shall be as you wish. Tomorrow we will change the director, we will change the professors, we will change the masters whom you don't favor.' Or if they demand champagne every day at dinner, then, all right, they will get champagne. We ought to know that the more one gives in to students, the more they demand and the more unreasonable they become. Our school must teach obedience and discipline, and if we can succeed only by force, then let it be by force."

Now he attempted to be matter-of-fact.

"At a meeting of the senior members we worked out a detailed plan of action. I shall explain our plan now and shall make every one of you responsible for the way it is carried out. I am sorry to say that we cannot count on M. de Guerle; his views, as you just heard, are very different from ours."

The director went toward the wall on which hung a large framed plan of Louis-le-Grand illuminated from both sides by candles. He felt like a general surveying his army of professors and masters. With a stick he pointed to the plan of his battlefield. Here he must beat the opposing army of rebels. And by God's grace and in the King's name he would win the fight.

2 : Tuesday, January 27, 1824

The bells of Louis-le-Grand rang repeatedly at five-thirty in the morning, until they beat away the hope of more sleep.

It was still dark when Evariste Galois awoke. He saw the familiar face of the master who was lighting a few candles in the candlesticks on the walls. Then he heard, "Up! Up! All of you, up!" and saw how the master flicked the covers from the beds of those who still lay upon them.

Evariste began to dress. He knew every detail of the room and every face in it. There were thirty-six beds, some of iron, some of wood, all placed exactly three feet apart. If these beds were to vanish, only the cold tiled floor and the small closets lining the walls would remain.

He looked up at the windows. They were hateful. They were so high that no one could reach them. When the light broke through, he would see the tip of a chimney and the depressing color of the winter sky. And then the crossed iron bars, forming small, dense squares! When he thought about Louis-le-Grand, these bars appeared first before his closed eyes. On a moonlit night their shadows spread on the floor, the beds, and the faces of his roommates. Looking at these walls, every morning and every night, he thought of prison. Did a prison look like this dormitory? It must be even worse than this.

The students dressed quickly in the cold room, vibrating with talk about the bugs that had bitten them during the night, the lessons they had or had not prepared, and with subdued, half-spoken allusions to the events that would come.

After dressing, Evariste went downstairs toward the lavatories. Their stench penetrated the building, increasing in intensity with the decreasing distance, until the air was so malodorous that breathing became difficult. In this foul air the students waited for an empty place, they rushed each other, those on the inside discussing with those on the outside the progress of their labors, the state in which they had found the place, and the state in which they would probably leave it.

When this was over, Evariste went back to his room, took a small towel, and ran with it to the fountain which stood at the center of the great court. Like other boys he wiped his face with the dry towel, put his hands under the fountain then dried them quickly, rushed back to his dormitory, put the small towel on a hook, grabbed a big Latin-French dictionary, Cicero's *De Amicitia,* Ovid's *Metamorphoses,* a notebook, and went to the study room for the fourth class. At six o'clock a master came in and the students began to prepare their lessons.

This was a pleasant time for Evariste. He would open Ovid's book, slightly moving his lips to convince the master that he was memorizing. The master, with sleepy, bored eyes would look apathetically for a victim who might try to talk to his neighbor. Evariste knew exactly what would happen during this one-and-a-half-hour interval of study. Just as on every other day, he would nurse thoughts and see pictures a thousand times more real than the world around him.

During this study hour he was never at Louis-le-Grand. He wandered just a few miles from Paris, but the two worlds, Bourg-la-Reine and Louis-le-Grand, were as far from each other as two worlds can be.

Evariste saw his father, so near and so distinct that he seemed to touch him. He felt his father's hand softly sliding over his head. When Evariste thought about his father, he also thought about light, about the sun that radiates warmth and melts snow, or about a clear day when the air smells of hay and flowers.

Smells! They express everything. Flowers and hay are Bourg-la-Reine. The sharp smell of urine is Louis-le-Grand.

His father laughed loudly. But recently the laugh often died suddenly, as though someone had cut it with a knife. Mother never tried to prolong Father's laugh. When Evariste thought about his mother, he also thought about a Greek goddess with black hair and shining black eyes. He smiled.

"Galois! You seem to be having a good time."

He heard the master's voice, but not his words, looked at Ovid's book and read mechanically:

Aurea prima sata est aetas, quae vindice nullo,
Sponte sua, sine lege fidem rectumque colebat.

How well he knew these verses! He could still hear his mother's cool, patient voice when she explained their meaning. He was free to think about Bourg-la-Reine because his mother had taught him all the Latin and all the Greek that he was studying now. Why had they sent him to Louis-le-Grand? Why didn't they teach him at home? His father and mother knew everything better than all his masters and professors, all of them taken together. Yes, these verses. He remembered how proud his mother was when he recited them well and smoothly in the home of his grandfather, M. Demante. He knew that his mother was proud though her face remained unchanged. But his father came to him, hugged and kissed him. Then his mother whispered to Father and his father's face clouded.

Then his grandfather asked him, "Evariste! What would you like to be when you grow up?"

Sometimes he thought that he would like to be an important judge like his grandfather. Sometimes he thought that he would like to be the mayor of Bourg-la-Reine like his father. What would he like to be now? Where would he like to be now? Away from Louis-le-Grand. But he must not hate Louis-le-Grand.

His father had told him, "You may hate ideas, but do not hate the men who represent them. Even if you could destroy these men you would still not destroy their ideas."

He would do his best not to hate the fat, red face of M. Berthot or the long nose and bad skin of the master.

When he had been at home a few days ago, he had told his father and mother that he disliked Louis-le-Grand.

His mother had said, "Without a good education you won't amount to anything. If you want to be a judge, or a doctor, or a scholar, you've got to go to school and earn degrees, whether you like it or not."

He had tried to argue, he had imitated his professors and the director and the masters until Father had begun to laugh aloud. Then Mother had brought everything to a sudden end.

"I hope you don't imitate us at school."

She had left the room. Evariste would have cried but for his father, who had talked to him then as to an equal.

"Everything you dislike at school is external. It won't matter much if you build your own internal life." He smiled weakly and seemed embarrassed when he added, "Look more into yourself and you will see less of Louis-le-Grand."

How often had he heard that ability isn't everything, that there is something more important. But this "something else" had different meanings for different people. This something meant obedience at Louis-le-Grand. It meant calmness and strength for Mother. What did it mean for Father? It was not easy to say. But it was connected, he knew, with the meaning of two words that he had heard so often: "liberty" and "tyranny." How differently these two words sounded, how differently the eyes of his father shone when he said each of them! They were as far apart as Bourg-la-Reine and Louis-le-Grand. Liberty was something for which one fought courageously and died gladly; tyranny was a force that, by the use of threats or a stick, compelled one to do hateful deeds. Liberty was light and tyranny was darkness. Liberty was Bourg-la-Reine and tyranny was Louis-le-Grand. But tyranny must be fought and Louis-le-Grand must be fought. They would fight it that day.

Why "they"? Why didn't he think, "Today *we* shall fight tyranny at Louis-le-Grand?" For them he was a strange newcomer who could not be trusted. For today they had not assigned any dangerous task to him. He would have to do only what the whole class would do: tear up books, set them afire, and beat up the master. The thought horrified him. The thought of throwing big dictionaries into the faces of grown-up men made him shudder. He wished the hour would never come. What would his mother say? Would she understand?

He heard a bell shrieking insistently. The bells! Sweet was the sound of bells in Bourg-la-Reine. Their melody was that of peace. The sound of bells at Louis-le-Grand brought discord and unrest.

Two men entered the room. They carried a large container of onion soup. Each student took a spoon and a bowl from the pile in the corner of the study room; then the waiter put a ladleful of soup in each bowl. In a few minutes it was eaten, the bowls removed, and the spilled fluid wiped with dirty rags.

Students who were only half-pensioners began to arrive. In their boardinghouses they had gone through the same routine as their comrades at the school. Then from eight o'clock in the morning until after dinner they formed one large community, and today one fighting force.

At eight o'clock, when the bell rang, M. Guyot entered the fourth class to face seventy adversaries. His back was slightly curved, his eyes were restless and tired. He opened the wooden gate in the fence surrounding the rostrum and sat on his chair, his face visible below a stone bust of Cicero.

Today the class was quiet. No tricks were played on M. Guyot, on whom trick-playing was so easy that it had ceased to be funny. Two weeks ago a rat had been thrown into the rostrum. But not today. Sometimes small balls of paper were thrown at Cicero's bust so that they recoiled onto M. Guyot's bald head. But not today.

Today the students submitted docilely to everything. They recited verses, translated, rephrased, analyzed, wrote exercises; they came one step nearer the goal set for them at Louis-le-Grand: to think and write in Latin. This was what distinguished a well-educated Frenchman.

But the external apathy covered an inner tension that grew hourly. The fourth class was proudly aware that it was the youngest in which the rebellion was organized. The older students had placed confidence in them, and they must not betray it.

At twelve o'clock the morning lectures ended. The students had a free hour in which to eat rice soup and a plate of meat with vegetables, to relax and gather strength for the afternoon sitting.

Evariste stood at the window looking out at the great court. He saw the gate open and a carriage being drawn in by two horses. Such entries were not unusual, but this was made so by the presence of the director and a few masters who shouted to the driver, telling him where to go and where to stop. Evariste changed his angle of vision and saw a second and third carriage, and even two horses' heads behind that.

Other students exclaimed joyously at the sight of the strange procession.

"Which are the masters and which are the horses? They look alike."

"Don't flatter the masters."

"What are the carriages doing here?"

"The masters are running away."

"They're scared."

"They know their time is up."

The professor entered the room. The students took their places, wandering slowly to their benches, looking impertinently at the professor of Greek, and saying with their eyes, "You wait! Only a few hours more! You'll see."

Evariste repeated to himself, "What is the meaning of these carriages? What are they doing in the court?"

He heard distant voices, then the rattle of a carriage leaving the court.

"What does it mean?"

Half an hour later the disturbing voices came back. This time they were slightly different, perhaps more violent; and again came the rattle of a carriage leaving the court. He tried to tell his neighbor about the noises and carriages, when he heard a commanding voice from the rostrum.

"Galois! Read the next sentence."

Evariste did not know what the next sentence was. He did not even know whether they were reading Xenophon or the New Testament in Greek. He stood up without saying a word. His neighbor tried to open the right book at the right place. Yes, it was Xenophon. But Galois stood motionless.

The professor now carried out the director's orders, "Behave as usual, pretend you know nothing." He turned toward Galois.

"Ha! I see! We were dreaming, weren't we? You have your private thoughts, much more important than all we are doing here."

The professor pedantically rounded out each word.

"Your thoughts are very important, I am sure. Perhaps you have solved some great universal problem. Tell us something about it. Let us all learn about your deep and penetrating thoughts."

The soft, ironical voice changed suddenly into a cry of anger:

"What were you doing?"

There was no reply.

"Being obstinate, ha! We shall remember that."

He scribbled something in his notebook and pronounced with authority, "You are lazy, inattentive, and talkative.

Then he turned to another student.

The lessons ended at four-thirty in the afternoon. The students left their class and went to the study room, where refreshments were waiting for them: pieces of stale bread sprinkled with water. With their mouths full they whispered to each other.

"Only one and a half hours more."

"Start right after we four give the signal."

"Watch for the bell."

"It'll be all right, if we stick together."

They had to talk in whispers. The professor was still with them, waiting for their master's arrival. This master! How well they knew his long horselike face with its tremendous nose; how humble and subdued he was when he talked to the professors, always avoiding their eyes. But the same face became arrogant and spiteful when he talked to the students. He spied upon them at night, reported when the slightest rule was broken, pedantically wrote down the name of a student when he came a minute late from his leave, abused and threatened the pupils in a sweet voice never raised in anger or indignation. Today, as every day, he would supervise their studies until six o'clock, but then he would be in their power. Yes, it would be a pleasure to give him one hell of a beating.

A master entered the study room, but his face was different from the one they expected to see. Where was the long nose and the pimply skin?

They listened to an energetic voice that excluded all opposition.

"Today I shall supervise your studies in place of M. Ragon, who is unable to come. You will do your last two exercises in Latin and Greek. Commence!"

Someone knocked at the door and opened it without waiting for an answer. It was the school janitor carrying a black book which he handed to the master.

"Foublon."

The student rose.

"Go immediately to the directors' study."

Foublon hesitated; there was complete silence while all eyes focused upon him.

"Did you hear what I just said?"

Foublon went out and the janitor closed the door behind him.

"Terrin."

The student stood up.

"Go immediately to the directors' study."

He left the class and the tense silence increased.

"Bouillier. Fargeau."

They left too.

The students looked at each other in surprise and fear. Something unexpected had happened. These were exactly the four students who were to start the rebellion. Would they return before six o'clock?

The tension grew. The students wrote notes to each other.

"Who'll start?"

"What about the new master?"

"How can we attack him?"

"Who'll start?"

"Will they be back in time?"

"If they don't come back in time, who'll start?"

Evariste's thoughts began to wander along a tangent which inevitably led to Bourg-la-Reine. But then he became more and more restless, more and more disturbed. Now, like all the others, he waited for the bells to ring, the signal to put out the candles placed between each boy and his neighbor. Protective darkness would save the individual from punishment and spread responsibility over all, rendering anonymous both guilt and heroism.

But the bells did not ring. The students, accustomed to the clocklike schedule, reacted like clocks. They knew in their bones, with their minds, with their skin, that six o'clock had passed.

"Who will start?"

"When do we start?"

"Who'll start?"

Evariste thought, "There won't be any bell. Of course, there won't be any. We think that bells are rung by time itself, that their voices come into the classrooms as day and night come into the world. But the bells are moved by human hands, and human hands can be stopped. The rules of Louis-le-Grand are as hard and rigid as steel. If Napoleon came out of his grave, if Paris burned to the ground, the bells of Louis-le-Grand would ring as they rang yesterday and a hundred years ago. But today the bells are silent. Their silence spreads confusion; it is their silence that will break the spirit of the revolt; their silence will increase fear and force submission."

The wave of his anger rose, his cheeks burned, he felt a pain in his eyes and heard the loud beating of his heart. He stood up. Every head turned toward Evariste. His face was red and his eyes were afire. He stood motionless, then opened his mouth and closed it again. His right hand moved along the table until it touched a fat, heavy book. He grasped the Greek-French dictionary, raised his right hand and threw the volume toward the candle on the master's desk. It was a good hit. Throwing the candle into a horizontal position, the dictionary put out the flame, passed the edge of the desk, and landed noisily on the floor. Someone threw another dictionary which just missed the candle on the other side of the master's table. From all sides dictionaries began to fly. Most of the candles placed between the students went out. Along the walls some candles still burned, throwing fantastic shadows in the dim classroom. The master stood up; with wobbling steps, facing the class, he moved toward the wall.

Someone shouted, "Down with the Jesuits!"

Another voice responded, "Down with Berthot!"

The chaotic class chorus repeated, "Down with the Jesuits! Down with the Jesuits! Down with Berthot!"

Suddenly and with a bang the door opened. Everyone turned and the shouting ceased. The master stopped his retreat; the students standing on the benches remained motionless.

Calmly and majestically the subdirector, M. Gustave Émond, walked toward the front of the room. He seemed neither surprised nor angry.

Facing the class he said softly, "I have come to make an important announcement."

Only then did he seem to notice the dark candles, the students standing on the benches, the dictionaries scattered on the floor.

He looked around carefully and said, "You may be seated."

There was magic in these words. The students sat down, some of them jumping off the benches, some of them returning to the places they had left. They all tried to make the transition to their normal state as silently and as quickly as they could. Evariste found himself doing what others did: looking in fascination at the man who had entered the room, obeying his commands, and listening to his every word.

"I came here to make an important announcement. I am sure that most of you, and perhaps all of you, will welcome what I have to tell you."

He paused and looked at the silent audience spellbound by his words.

"For a long time we have known that some of the students were trying their best to spread discontent and dissatisfaction among you. They did it unscrupulously by deceiving you, by spreading lies and false rumors. They spread rumors that the school would be returned to the Jesuits and that such is the intention of our director, M. Berthot. I do not need to tell you that they lied to you and lied very stupidly. The students who spread these rumors knew perfectly well that they were lies. But they counted, I am sure unjustly, on the stupidity of their fellow students. They appealed to your friendship and to your feeling of loyalty. They tried to persuade you, but they were ready to use force and terror if necessary. They tried to endanger you, and they planned their actions in such a way that you and not they alone would be punished.

"Fortunately, I can tell you something that you will be happy to hear."

Suddenly and dramatically he raised his voice.

"These students, forty of them, were expelled from Louis-le-Grand today."

It was so quiet that against the dark background of calm, the sizzling noise of the few burning candles seemed unbearably loud. Forty students, the best students, were expelled from Louis-le-Grand, torn out from the midst of their comrades to face the wrath or despair of their parents. The dramatic voice paused just long enough for the most stupid student to understand what the sentence meant; that what had happened here in the fourth class had happened in every class at Louis-le-Grand. Now the students remembered the carriages. Now they knew that the strange noises meant cries of resistance; each rattle meant that one more leader of the planned rebellion had left the school forever.

The dominating voice continued, "These students will never return to Louis-le-Grand. Probably all opportunity for education in France will be closed to them. You are now free of their terrorizing. You can now pursue your studies peacefully."

The voice descended from the high tones of drama to a composed calm.

"We should like to regard the whole incident as closed. Even if some of you are guilty of negligence, of not telling the authorities about the plan of the revolt, we wish to forget it and continue with our normal work. You were sent here for this work and we, your teachers, are responsible for it. You will understand that to do this we must be sure of your loyalty, we must be sure that you do not feel bound by any promises you may have made to the rebellious students. Otherwise it is obvious that if you share their views, you ought to share their fate. You must agree with me that this is both logical and just."

He looked around to see whether there was anyone who thought the argument illogical or unjust. No one broke the silence.

"I wish, as I am sure you all wish, to finish this painful incident and to forget it. But first I want you to pledge loyalty to our school; I want you to tell me that you do not feel yourselves bound by any promises you may have given, willingly or unwillingly. I shall read your names, one after another, and those of you who are for order, discipline, and loyalty to our school will make it clear by saying 'I promise.' Of course you understand that there is not the slightest compulsion to give me your promise. You must do it of your own free will. Otherwise it is worthless. I shall now read the list.

"Adelier."

A thin boy stood up and in a trembling, soft voice whispered, "I promise."

"You must say it louder so that we can all hear you. And don't do it unless you wish to."

A louder voice mingled with tears repeated, "I promise."

"That's better."

Evariste felt the icy cold in his fingers and the burning heat on his cheeks and forehead. He whispered to himself through stiffly closed lips, "I promise! I promise! I promise you that I shall never forget this to the end of my life. I promise that I shall never forget this great lesson in perfidy and hypocrisy. I hate you and all men like you! You have taught me to understand what hate means. My father tried to teach me that we can live without hate. Not here, not at Louis-le-Grand. I shall always hate men like you; men who suppress those who are weak. I shall always fight men like you whenever and wherever I meet them. I promise! I swear to God and with all my heart. I promise. . . ."

Evariste heard a quiet and indifferent voice.

"Galois."

He stood up. A shout swelled by suffering, anger, and passion crystallized in the words, "Yes. I promise."

M. Émond looked up at Galois and saw a young triangular face, broad at the forehead and forming an acute angle at the chin. The eyes were deeply set, and their gaze seemed to penetrate to the inside of the objects on which they focused. M. Émond looked away with an effort, and before calling the next name he murmured to himself, "That is a very strange boy."

3 : Wednesday, January 28, 1824

Every year on Saint Charlemagne's Day the best students, selected long beforehand, attended a banquet during which torrents of Latin and flowery French oratory were poured out both by the professors and the students.

The Saint Charlemagne banquet on Wednesday, January 28, 1824, was unlike any other in the long history of Louis-le-Grand. Some weeks before, one hundred and fifteen students had been selected for the banquet, but the day before, forty of these very students had been sent home.

The great dining hall was brightly lighted. It was adorned with ferns and flowers. White flags with the lilies of the Bourbons hung on the wall behind the professors' rostrum with its long table. The students' tables were at right angles to the master's table, which stood on a raised platform.

The seventy-five students came in silently. They were dressed in their Sunday blue uniforms. They looked at the empty plates before them and at the forty empty places among them. As the procession of teachers headed by M. Berthot entered the room, they stood up, respectfully fixing their eyes on the floor. Then they sat down humbly, beaten dogs that had learned their lesson.

The director looked triumphantly at the students below. Some of the students raised their eyes and saw the beaming director and the five subdirectors sitting beside him. They looked for M. de Guerle, the proctor, the man they loved and trusted. But he was not there; he had not wanted to witness the humiliation of his pupils. Not one friendly face there above, not one softened by sympathy or pity.

No word was spoken on the rostrum where the professors sat and no word was spoken among the students. The silence was interrupted only by the noisy drinking of soup, the impact of knives and forks on the plates as chicken was carved; even the chewing could be heard in the silence. The room seemed dark and gloomy although all the candles were lighted. The dessert was accepted without comment and eaten without haste. Even the champagne was received with apathy and indifference.

The oppressive atmosphere created the silence, and the silence thickened the oppressive air which nurtured the tension. It was worse than a funereal atmosphere.

The director rose. With his fat hand he took a glass of champagne, smoothing his short beard with the other.

"I drink to the health of our beloved King, Louis XVIII."

And the unexpected, the fantastic did happen!

The students looked at each other. They knew what was expected of them. They could burn their school to the ground, they could beat their master, but they could not refuse to drink the King's health. They had not conspired to refuse this toast. But not one of them was willing to stand up first. They looked at one another with a defiant look, challenging any who might dare to rise. But no one did. They all sat, while the standing director and the professors gazed in horror and amazement at the wax figures rigidly stuck to their benches and boldly returning their masters' gaze. The humility of their submission was gone, defeat was changed into triumph. Theirs was the revenge now. They stared at the director's scarlet face. They saw him and the professors subside into their chairs, trying to appear aloof and indifferent. Silence returned to the hall, but it was a different silence now. The feeling of triumph had left the high table and descended to take its place among the students. The feeling of humiliation had ascended the rostrum to take its place among the professors.

M. Émond looked enquiringly at M. Berthot, who slowly nodded his red head. M. Émond rose. The magic voice would be heard. But now small drops of perspiration peppered his forehead. So this godalmighty who had descended from Olympus upon Louis-le-Grand was afraid and perspiring. He raised his glass.

"I drink to our director, M. Berthot."

The magic voice had lost its magic. Not one of the students moved. Only their eyes glowed still brighter as they looked with amusement at the mummies on the rostrum, too proud to show their embarrassment and too clumsy to conceal their despair. Like marionettes they repeated "long live," which now sounded more like a mockery than a wish.

A young professor blushed, rose, and recited quickly, "I drink to our absent proctor, M. de Guerle."

This time there was an explosion, a sudden burst of long-restrained energy.

"Long live M. de Guerle!"

Some of the students stood on the benches, shouting louder and louder with all their force. They became hysterical. They repeated the same words again and again in chorus:

"Long live M. de Guerle!"

One of them ascended from the bench to the table, kicked some plates with their pieces of chicken onto the floor, kicked away a few glasses of champagne, and, swinging a fork, led the chorus, giving rhythm, power, and uniformity to the repetitious phrase. Other students picked up forks and spoons to beat the rhythm of their cry on the plates and glasses, smashing them into pieces and spilling wine on the floor.

The director pounded his fist on the table.

"Silence; silence. I want to tell you something."

But whenever his words became audible, the cry, "Long live M. de Guerle!" increased until it swallowed up the director's shouts. His agitated mouth and swinging fists were visible, unrelated to any sound.

Finally his words broke through the armor of noise.

"Silence! Silence! I want to tell you something. You are no longer students of our school. You are no longer our students; you are not our responsibility. You are expelled from Louis-le-Grand. You are going back to your parents, all of you. Silence! I repeat. . . ."

The hysterical cries ceased and there was silence. No one tried to revive the cry again; its words were played out and empty. Something different must come now. Even though they must leave school, they would not be defeated. They waited for leadership, for someone to show how to manifest the strength they had discovered in themselves that day.

The silence was broken by a clear voice singing the first words of the banned Marseillaise. The song swelled and grew. It grew in volume and it grew in emotion. Their fathers' fighting song, buried deep in their hearts, was with them again. They sang the words that had lit the fires of liberty, the melody of struggle, victory, and of a glorious France, the words and the melody which some of them would sing six years later, fighting and dying on the streets of Paris.

4 : September, 1824

The death agony of Louis XVIII, the last French King to die on French soil, had lasted for three full days. It was witnessed by a crowd of courtiers in the stifling heat and silence broken only by groans from the suffering man. Before he died, he raised his white hand with its gnarled and immobile fingers over the head of the three-year-old Duc de Bordeaux and whispered, "God bless you. May my brother guard the crown tenderly for this child."

Mme. du Cayla obliged her protectors by persuading the dying King to call a confessor. For this she received eight hundred thousand francs.

On September 16, toward four o'clock in the morning, the gentleman who was holding the bed curtain let it fall and announced that the King had "ceased to breathe."

Nine days later, the coffin of Louis XVIII was lowered into the vault of the Cathedral of Saint Denis, and for the last time these gloomy walls received the body of a king of France.

A row of heralds threw upon the coffin their caps and coats of arms, and each time this tragic gesture was accompanied by cries, "The King is dead! The King is dead!"

Three dukes came forward. Each of them threw into the vault the colors of the Royal Guard that he commanded and each time the heralds repeated, "The King is dead."

Then the crown, the scepter, the spurs, the breastplate, the sword, the shield—all the war insignia of this most unwarlike king—were thrown in, and the cathedral resounded with the ring of metal and the heralds' cries, "The King is dead."

The Grand Chamberlain, the Prince de Talleyrand, limped forward and lowered the standard of France over the coffin.

Then the Master of the Household came forward and rapped his heavy stick three times on the stone floor. When the hollow sound died away he shouted, "The King is dead, the King is dead, the King is dead; let us pray for the soul of the dead King."

All heads lowered in silence.

The Master of the Household rapped his stick again.

"Long live the King!"

The door to the sepulcher closed with a crash, the drums beat, the trumpets sounded and the chorus of heralds recited:

"Long live King Charles, tenth of the name, by the Grace of God, King of France and Navarre, most Christian, most august, most powerful, our most honored lord and good master, to whom may God grant a very long and a very happy life. Let all shout 'Long live the King.'"

Thus the reign of Charles X, last legitimate Bourbon king of France, had begun.

III

"I AM A MATHEMATICIAN"

1 : May 29, 1825

IN JANUARY, 1825, the *Moniteur* announced that the coronation of Charles X would take place in Rheims that spring. The citizens of Rheims looked with pride and gratitude at the towers of their cathedral from which a shower of gold would descend upon the town. Before long even a dark hole, if it contained a bed, rented for sixty francs a night.

The British Ambassador, Lord Northumberland, sent his steward to find lodgings in Rheims. He saw a "For Sale" sign before a large house and asked the owner, "How much?"

"Ten thousand francs."

"I only want to rent it."

"For how long?"

"For the three coronation days."

"Then it will cost thirty thousand francs."

A month before the coronation, a swarm of masons smashed all loose pieces on the cathedral's sculpture, for fear one might drop on the King's head. Fragments of Christ's face and pieces of angels' wings found their way into the rubbish.

In May, the *Moniteur* announced joyfully that the King would be anointed with the ancient holy oil that had been brought down by a dove from heaven. The precious ampulla had been preserved in Rheims for centuries. But in the terrible year 1793, citizen Ruhl, a representative of the people and a commissioner of the convention, grabbed the holy bottle from the cathedral, broke it upon the head of Louis XV's statue, spilling the oil over the stone king and the mud below. Yet a miracle occurred, and some trustworthy though unnamed personages collected the holy drops from the stone and mud, keeping them in trust for the great day when a Bourbon king would again be crowned in Rheims.

The procession entered the cathedral early in the morning of the coronation. The King was attired in a cherry-colored gown striped with gold, and the peers of France who surrounded the King wore long mantles of velvet and ermine embroidered with gold.

Inside, the austere Gothic cathedral was altered for the occasion into a Greek theater, and the performance took place under a canopy made of crimson satin. The Archbishop and the King were leading actors in a play that took five hours and in which the King changed his costume six times. In this he was aided by his cousin, the first prince of the blood, Louis Philippe, Duc d'Orléans. In one scene the King lay prostrate on cushions, his handsome face and gray hair touching the carpet on which the Archbishop trod. The representative of the Holy Father pricked the King's flesh with a golden needle through seven holes made in Charles' clothes. In another scene the King knelt before the Archbishop, after receiving the scepter in his right hand and the symbols of justice in his left hand. And the Archbishop put first the miraculously-preserved holy oil and then Charlemagne's crown upon the King's head.

Some among the audience remembered a very different, though equally colorful, spectacle that had delighted them some twenty years ago. It was performed not in Rheims, but in the Cathedral of Notre Dame in Paris. It was then not the Archbishop, but the Pope himself who came from Rome to crown the young god of war. And never did Napoleon lie prostrate before the Holy Father. No! His Holiness was not even allowed to touch the crown. It was Bonaparte himself who grasped Charlemagne's crown in his own imperial hands and put it firmly upon his own imperial head.

Those among the audience who hated the ultras watched the spectacle in fear that soon the King would swear the old oath of French kings: to preserve the rights of the Church and to exterminate heretics. With relief, they listened to the new words in the ancient rite: the King swore to obey the constitutional charter.

When finally the ceremony ended and Charles in all his kingly regalia sat rigidly on the throne, the worn-out spectators shouted, *"Vivat rex in aeternum."*

The Revolution? The Empire? These were only short, dark episodes in France's glorious past. Now, when the Bourbon line was to reign forever, the traces of these days must vanish from the land and their memories must die in the hearts of men.

The great curtain that cut off the end of the cathedral was drawn aside, the crowd rushed in, the bells pealed, the organ played, trumpets blared, cannon shots replied to musketry fire, and hundreds of doves were released from the vaulted roof; they fluttered in a cloud of incense, terrified by the noise of the multitude.

Thus the last French king was crowned in Rheims.

2 : 1825–1827

These were the years in which the beaten French bourgeoisie began to raise its head again, coining two slogans for its fight with the ultras. The first, *vive la charte,* had little effect. The people did not wish to worry about the charter, which the King swore to respect. The second slogan caught the imagination of France and inflamed the nation. Through the length and breadth of the land it was repeated over and over again in varying words: "Down with the Jesuits." "Down with the congregation." "Down with the black priests." A liberal paper philosophized, "Our age will be difficult to explain to our children. Theological controversy is the order of the day and we hear of nothing but monks and Jesuits."

The stupidity and blindness of the pro-Charles and pro-Jesuit ultras was the opposition's best ally. The liberals repeated and repeated the same arguments: France was governed by the King, but the King was a marionette in the hands of the Jesuits. The chambers passed a law that theft of sacred objects from the church was punishable by death. The same chambers passed a law that punished profanation of the Host on a par with parricide. The King had prostrated himself before the Archbishop at Rheims. Did it not show how the Jesuits wished to turn the clock of history back to the Middle Ages and to the times of inquisition?

Soon the shopkeepers and tradesmen of Paris saw a performance even more alarming. At the feast of the Church, at the Jubilee celebration in 1826, religious processions marched through the streets of Paris and the King appeared in all of them. In the last and the most pompous one, the foundation stone was blessed for the monument of the martyr-king, Louis XVI. Charles X, members of the royal family, cardinals, bishops, two thousand priests, marshals, generals, staff officers, peers, deputies, civil officials, magistrates, formed a procession that in length and magnificence surpassed all others.

Artillery roared when the procession arrived at the Place Louis XV. The Archbishop of Paris mounted the steps of the great altar. Three times he called to heaven for mercy and forgiveness, as all present fell on their knees. Then the King, dressed in a violet robe, the color of royal mourning, came forward to lay the foundation stone which the Archbishop would bless. To the Parisians ready to watch any colorful performance, these two robes—the King's and the Archbishop's—looked very much alike. Hardly had the procession returned to Notre Dame to the thunder of guns, through streets lined with troops, when a new rumor began to radiate from Paris: that the King was made a bishop, that he was a member of the Jesuit Order, that the procession was a penance imposed by the Church upon him in atonement for the errors of his youth. Wilder and wilder rumors followed through spoken words and printed pamphlets: that no one would be able to secure an office unless he was a Jesuit; that the priests could raise a thousand hands well armed with daggers; that the Pope could depose a sovereign if he wished to do so. The rule of fanaticism was depicted as more dangerous to the fields, workshops, and factories than the anarchy of the most bloody revolutionaries. The government tried to stop these arguments by dragging before the courts those who "threw contempt upon persons or things connected with religion." Yet in most cases the accused were freed by the judges, after which their language became still more violent and abusive. The Parisian shop windows displayed pictures of priests with extended stomachs and obscene faces, or drawings of thin ascetic monks burning Voltaire's books. The real Jesuit bogey was blown up to unreal dimensions until it threw a shadow of hatred and fear over the whole of France.

In cafés, clubs, wine shops, the most frequent word was "Jesuits." In Royal Colleges the students repeated what they heard from their parents. The mood that prevailed at Louis-le-Grand is best pictured by the dismayed letter that M. Laborie, the new director, wrote to his superior, the Minister of Education:

> There is no religious spirit among the students. The very few who are pious are ashamed to make the sign of the cross, fearing that others would respond with sarcasm and laughter. To them nothing is sacred. Their spirits and hearts are savage. Wickedness is at its peak here, and there is little hope for improvement. Even the professors give a bad example as they fail to go regularly to chapel. The parents give a bad example, exciting the imagination of their sons and inspiring them to revolt by incessant talk about the famous order of the Jesuits and the dangers of church domination. The Jesuits are the most popular topic among the students. How can we deal with youths convinced that their rebellious action will have the approval of their parents?

After the rebellion at Louis-le-Grand, M. Berthot, the disgraced director, had been dismissed. He had thrown out the hundred and fifteen best students, the school's pride and flower, all who had won competitions and made Louis-le-Grand the most distinguished among the Royal Colleges. Yet, and this he was not forgiven, the rebellious spirit prevailed. Thus M. Berthot was removed and his successor M. Laborie appointed. M. Berthot had been brutal, crude, clumsy. M. Laborie was well educated, skilled in intrigue, and he loved the King. The same spirit would direct the college, but the hand would now be gloved to soften the pain and velvet the noise when strangling students' rebellions.

These were the years when Evariste Galois progressed step by step until he reached the rhetoric class. He never forgot that only his absence from the Saint Charlemagne banquet had saved him from the fate of a hundred and fifteen dismissed students.

These were the years when the teachers at Louis-le-Grand complained that the student Evariste Galois was dreamy, lacking in discipline and ambition. He may have ability, they said, even remarkable ability, but he is immature and peculiar. The director strongly advised M. Nicolas Gabriel Galois to let his son Evariste repeat the second class. But old M. Galois disagreed and thus it happened that, in the autumn of 1826, Evariste entered the rhetoric class, counting in days the diminishing distance from freedom.

3 : February, 1827

A sharp "Come in" answered Evariste's knock; he entered the study and stood at the door while the director continued writing. He looked at the director's sharp features, the tightly closed lips, and the thin ascetic face. Then he looked at the desk, counted all the objects on it, looked at all the pictures on the walls, then again at the director.

Evariste thought, "You know very well that I am waiting here. It is a new kind of torture invented by the great master of inquisition, M. Laborie, the director of Louis-le-Grand.

"I shall remind you of my presence. I shall come to you very, very calmly, then suddenly I shall grab the pen from your hand, break it, and say, 'We all hate and despise you. You are a Jesuit, a Jesuit, a Jesuit of the short robe.' Will you notice me then?"

The director looked up.

"Oh, yes, Galois."

He put aside his pen, leaned back, and spoke from far above, very slowly and very distinctly.

"Galois! I have read and discussed your record. It is not what we all expected."

Galois replied, but only in his thoughts, for the school had taught him to keep his thoughts to himself, "That's because I don't like you or the school. I knew you wouldn't let me finish school this year no matter what I did or how I worked. These were your orders from the black priests."

M. Laborie waited for an answer, but it did not come.

"We thought you were too young for the rhetoric class. You're not yet sixteen. But we thought we might be wrong and we didn't want to insist. Unfortunately for you, and against our hopes, time has proved that we were right.

"We are sure that you will be much happier in the second class. You'll be in the division of an excellent man, M. Girardin, you will meet boys of your own age, you'll find the work much easier, and your progress will undoubtedly be much greater."

The director paused as though waiting for an answer and then began to drop his well-polished words:

"All we care for is the good of our students. Because of this, we try not only to give you knowledge and to develop your minds, but, above all, we try to build your characters. This you will learn to appreciate when you are older. It is too much to hope that you can do so now. But staying at Louis-le-Grand one year longer may open your eyes. You will not only gain knowledge, but, what is much more important, you will gain maturity and understanding."

Again there was no reaction. M. Laborie turned toward Evariste.

"Do you understand what I'm saying to you?"

"Perfectly, sir."

"Then do you agree with me?"

Galois did not answer.

The director repeated in a voice in which there was not the slightest trace of impatience or annoyance, "I asked you whether you agree with me."

Evariste suppressed his growing anger and succeeded in saying calmly, "No, sir."

The director looked at him with interest, and the friendly voice sweetened still more.

"Tell me, then, why you disagree. Perhaps by discussion we may hit upon a solution which will satisfy us both. We may be able to find such a solution easily if what you suggest is good for you. Our interests don't contradict one another; indeed they strengthen each other. Tell me then, Galois, why aren't you convinced by my arguments?"

Evariste felt the storm coming, the words of abuse and violence pressing into his mouth. He knew that soon he would not be able to resist their growing pressure. They would pour out and beat upon the ears of the thin, ascetic head. He searched in despair for thoughts that could calm the storm and force the words to remain unsaid.

He thought about his father. He would have to repeat to his father exactly what the director said, and what he replied. He must behave so that his father's eyes would not become sad and clouded. Something was happening to his father. It was a long time since he had seen him gay, composing rhymes, imitating his friends, and laughing in a way that affected all of them—all of them, that is, except his mother. There must be some reason for the sudden change. Whatever it was, he mustn't add new reasons. He was now his father's spokesman.

Evariste said, humbly, "Sir, I wonder whether it would not be better for me to remain in the rhetoric class. I hope I shall be able to finish it successfully. And if not, I shall be ready to repeat the rhetoric class next year."

M. Laborie looked at Galois as though he had expressed an excellent idea of which the director had never thought before.

"Let us discuss your plan dispassionately and see which of the two plans is better for the school and therefore also for you. We want you to finish our school with a good record. We want to be proud of you, but we also want you to be proud of Louis-le-Grand.

"If you go back to the second class, where you were not bad before, you have a very good chance of taking part in the general competition and—who knows—you may win a prize. Then, with such preparation, you will have an equally good chance next year in rhetoric. But, if you stay in rhetoric, then you may just pass and even that I doubt very much. I am almost certain that you will have to repeat your last year starting with a bad record; whereas, going back to the second class, you will start your last year with a good, perhaps a very good, record. The more I think about it, the more I see that our plan is much better both for the school and for you. Yes, I am now quite convinced that ours is the better plan."

Then he turned toward Evariste with an air of finality.

"I hope that by now I have convinced you."

"I must end this conversation, I must end it at all costs. If I stay here any longer I shall spit at this Jesuit face. I must end it, end it now."

Evariste said meekly, "Yes! I am convinced."

And he felt like spitting upon himself.

4 : 1827

Evariste went back to the second class, to the old lectures, the old boredom amid new classmates.

Dreading the monotonous repetition of an old familiar curriculum, Evariste decided—for the first time—to take mathematics. It was an unpopular subject among the students and not considered important enough by the faculty to be made compulsory. As a result, a heterogeneous group of students from the third, second, and rhetoric classes met four hours a week to plow through the elements of geometry. When Evariste entered this class for the third term, the students had been exposed to about half of *Eléments de géométrie,* written by the great mathematician Adrien Marie Legendre, a book that was to influence textbooks on geometry for years to come.

During the first preparatory hour Evariste opened Legendre's volume and read the first sentences:

I. The object of the Science of Geometry is the measurement of space. Space has three dimensions: length, width and height.

II. A *line* is length without width.
The ends of a line are called points; a point has no extension.

III. A straight line is the shortest path from one point to another.

IV. Every line that is neither straight, nor composed of straight lines, is a curved line.

The next sentence referred to a drawing. The drawings did not interrupt the text but were collected at the end. Evariste unfolded the first sheet of drawings, read the text, and looked at the corresponding figure. He then went quickly through many definitions and came to the next section, which began:

"An axiom is a proposition that is self-evident."

He thought, "What is self-evident? What is self-evident for one may not be self-evident for another. Can something be so self-evident that it is self-evident to everyone, always?" He read:

A *theorem* is a truth which becomes evident by means of reasoning called *demonstration*.

He thought, "Then geometry deals with truth. There are theorems which are true. What we achieve by reasoning is to make the truth of these theorems evident. But, of course, their truth can only be as evident as that of the axioms on which they are based. The whole structure of geometry is based on axioms. What are these axioms?" He found the answer when he turned the page:

AXIOMS

1. Two quantities equal to a third are equal to each other.
2. The whole is greater than any of its parts.
3. The whole equals the sum of the parts into which it is divided.

4. There is only one line connecting two points.

5. Two lines, surfaces, or solids are equal if, when placed one upon the other, they coincide in all their dimensions.

As he read page after page, he saw the building of geometry erected with the simplicity and beauty of a Greek temple. Reading quickly, he saw not only the particular theorems, but their interrelation, the architecture of the whole, and the magnificence of the structure of geometry. He found himself anticipating and guessing what was coming next; he saw the structure increasing before his eyes. Soon the class, his surroundings, comrades, masters, noises, smells, ceased to exist. Abstract theorems of geometry became more real than the world of matter. The building of geometry grew in his brain. While reading the theorems, he nearly always saw in one flash how to prove them and then glanced quickly through the text and drawings for confirmation of his thoughts. Soon he could omit the proofs; soon he anticipated many theorems and felt that he had known geometry for a long, long time. But his knowledge had been hidden from his consciousness by a dark curtain. The reading of Legendre's book tore aside the curtain and revealed the Greek temple. He felt as though a strong, helpful hand had removed him from Louis-le-Grand; he was no longer unhappy, because Louis-le-Grand had ceased to exist for him.

During other lessons, during every free moment of the day he read, absorbing the theorems, making them evident by his own proofs, by his own reasoning. On the day he started to read Legendre, he reached "Book IV" about regular polygons and circles, and arrived at the problem: Find a circle which differs as little as one may wish from a given regular polygon.

He thought, "What kind of number is π?"

For an answer he turned to the notes in small print designed for advanced students, and there he found a proof that the ratio of the circumference to the diameter, and also the square of this ratio, are irrational numbers. Here the reading became more difficult. He met new symbols like tan *x,* the meaning of which he did not know. He turned to the last part of Legendre's book, to the *Traité de trigonométrie,* where this and other trigonometric symbols were defined.

When the lights went out in all the dormitories at nine fifteen in the evening, Evariste lay on his bed with open eyes, staring into space. He *saw* all the theorems he had learned during the day. Geometrical figures appeared, crossed out by equations spreading in all directions. A new theorem demanded that he make it evident by demonstrating its truth. The world of reasoning and the world of dreams formed a fantastic mixture of logic and imagination in which people resembled formulas and theorems resembled living creatures. Evariste tried to keep the two worlds apart, but he could not prevent their union through an exhilarating and restless night.

The next morning he read Legendre again. For the first time since he had come to Louis-le-Grand, he did not think about his father, he did not smell the hay, and he did not hear the ringing of the bells at Bourg-la-Reine. His mind burned with a new flame which death alone could extinguish. In two days he had finished Legendre's book designed for two years of study. He knew everything in it, and he knew that all he had grasped would remain and grow in his mind to the last day of his life.

During the lesson in mathematics Professor Vernier turned to Evariste.

"You are a new student in this class."

Evariste stood up. M. Vernier's eyes were tired and friendly.

"This is a new subject for you. You may find it difficult at the beginning. It will take you some time to get used to it. I shall leave you, say, a month's time before I examine you."

Galois stood silently staring at the professor's face. M. Vernier now looked impatiently at him.

"Do you think that you will be able to do it in a month?"

"Yes, sir."

M. Vernier began his lesson. It was about regular polygons which can be inscribed in, or circumscribed about, a circle. Most of the students seemed bored. The teacher's voice was dull and colorless. He repeated the theorems in the same form as they appeared in Legendre's book; he proved them using the same notation and the same arguments. He watered the reasoning by adding new sentences and by repeating them many times. The teacher copied the drawings from the book onto the blackboard, and the students copied them from the blackboard into their notebooks. When asked questions, they repeated the sentences which they had heard from the teacher and which in turn were those printed in Legendre. Most of the students learned these propositions as one learns poems in Latin or Greek, by repeating them dogmatically without trying to unveil their meaning.

Evariste saw how the soul of geometry was tortured here until it became a lifeless skeleton, a collection of boring, meaningless sentences, memorized from one day to another. He saw how this school managed with unsurpassed skill to change beauty into boredom, logic and reasoning into dogma, a Greek temple into a heap of stones.

The library of the college was in a state of disintegration. The windows did not close, the light was bad, the walls and books were damp, and only a very few students made use of the library containing many valuable volumes in Latin, Greek, and history, but only a handful of mathematical books. When Evariste took out *Résolution des équations numériques* by Lagrange, the librarian tried to be funny.

"You know the rule; books can be kept only eight days. Do you expect to finish it in eight days?"

"I shall try."

He read the definition of algebra in the introduction:

Algebra, as usually understood, is the art of determining unknown quantities as functions of known quantities or those assumed to be known; and also is the art of finding a general solution of equations. Such a solution consists in finding, for all the equations of the same degree, such functions of the coefficients of the algebraic equations that represent all its roots.

Up to the present, this problem can be regarded as solved only for equations of the first, second, third and fourth degrees. . . .

He read Lagrange's book less quickly than Legendre's. His feelings were mixed. Exciting as he found the great work, there was also a feeling of dissatisfaction, even of disappointment, that increased with the number of pages he turned. In geometry he had clearly seen the structure, but here he did not. And he knew that he didn't see it because it wasn't there. The building of geometry had style, harmony, and beauty. Algebra was a strange collection of buildings of different styles, most of them just begun, none of them finished. Behind the haphazard collection one could not sense the mind of a great architect.

He tried to formulate the reason for his dissatisfaction. He thought about the fundamental problem of algebra: that of solving algebraic equations.

Algebra—that is, elementary algebra—arose from this very problem, and its beginnings lie in distant times. Modern algebra, the algebra of today, a great field of contemporary research, arose from this very problem, too, and its beginnings lie in the work of Galois.

Thus, to solve an equation may be an easy task known to antiquity, or a difficult task accomplished in the time of the Renaissance, or it may be, in a certain sense, as recognized by Abel and Galois, an *impossible* task.

To say $2x-1$ equals zero, if x equals $\frac{1}{2}$, means to solve a trivial equation that hardly deserves to be dignified by that name. From there we may go one step higher, to an equation of the second degree, like $x^2-5x+6=0$. Here we also ask for a number (or numbers) that, substituted for x, will satisfy this equation or, as we say, we wish to find the roots of this equation. Indeed put into the equation the number 2 or the number 3 in place of *x*, and you will see that each of these numbers does satisfy the equation $x^2-5x+6=0$. (x^2 means x times x; $5x$ means 5 times x.)

Even the study of these comparatively simple equations of the second degree led to a far-reaching discovery: that of imaginary and complex numbers.

One could easily argue, "This is a thin web of abstract thoughts, of speculative problems far removed from our ordinary life." Yet the equation of the second degree leads to complex numbers, and the complex numbers are the everyday mathematical tool of the engineers and physicists. Out of the dreams of the mathematician, out of the abstract web of his thoughts, modern science and modern technique were born.

In the equation $2x-1=0$, the numbers 2 and -1 are the coefficients. We find the solution of this very simple equation by dividing "one" by "two." Similarly in the equation $x^2-5x+6=0$, the numbers 1, -5, 6 are the coefficients. We can find the roots of this equation by performing some prescribed operations on these coefficients. Indeed we remember that the roots were 2 and 3. Now these numbers 2 and 3 can be found by operations prescribed in these two simple formulas: $2=\frac{5-\sqrt{5 \cdot 5-4 \cdot 6}}{2}$ and $3=\frac{5+\sqrt{5 \cdot 5-4 \cdot 6}}{2}$

Such prescriptions can be carried out if we know the coefficients on which to operate. In the case of an equation of the second degree these prescriptions are still simple, though much more complicated than for an equation of the first degree.

Some algebraic equations are *solvable by radicals.* This means that we can find their solutions by a finite number of operations performed on the coefficients of the algebraic equations. These are rational operations (addition, subtraction, multiplication, division) and extractions of roots. If a solution gained by only these operations exists, we say that the equation is solvable by radicals.

The solution of an equation of the first degree is trivial. The solution of an equation of the second degree is very easy. With the solution of an equation of the third degree, complications arise. But it can be done, and it was done almost three hundred years before Galois was born. We can find the roots—that is, the solution—of an equation of the third degree by methods familiar to every mathematician; the problem can be reduced to a known one, to that of solving an equation of the second degree. This is the method used over and over again in mathematics: to reduce the solution of a new problem to that of an old one, whose solution is known. Similarly, an algebraic equation of the fourth degree is solvable by radicals. For we can reduce the problem of its solution to that of the solution of an algebraic equation of the third degree, which is known.

But here the method that Lagrange explains in his book breaks down suddenly, completely, and unexpectedly. It is true that if we can solve an equation of the second degree, then we can also solve an equation of the third degree. If we can solve an equation of the third degree, then we can also solve an equation of the fourth degree. It would seem that this chain could be prolonged; that if we can solve an equation of the fourth degree, then we should be able to solve an equation of the fifth degree. Just as on a ladder, we should be able to climb higher and higher toward the solution of equations of higher and higher degrees.

Is it possible to climb from one equation to another, to reduce the solution of an equation of higher degree to that of the next lower degree? Is it possible to solve all algebraic equations by rational operations and those involving radicals? Or, in other words, can the ladder be prolonged indefinitely, or does it break down?

Galois felt that this was the most essential problem of algebra, a problem of which Lagrange did not know the solution. The method developed by Lagrange worked, up to the equations of the fourth degree, but for an equation of the fifth degree it led to an equation of the sixth degree. Thus a solution of a problem was "reduced" to that of a much more complicated one. It was like learning to jump from the roof of Louis-le-Grand by practicing from the tower of Notre

Dame. If, again, Lagrange's method was used to solve an equation of the sixth degree, then the problem was reduced to that of solving an equation of the tenth degree. It was like trying to reach the tower of Notre Dame not by climbing, but by jumping at it from the peak of Mont Blanc!

At first Galois believed that a method must exist by which all algebraic equations could be proved to be solvable by radicals. And it is not important whether it could be easily done in practice or not. But to find a proof that it can be done, that such a solution always exists, seemed to him the central problem of algebra.

It was only a few weeks after Galois had read Legendre's geometry that he began to formulate his own problems. He was less than sixteen years old, yet he had experienced both the suffering that comes from groping in the dark, and the ecstasy of understanding. The world around him became shadowy. The school, his teachers and comrades, all became unimportant, almost nonexistent. By abstract thought he built around himself an impregnable wall through which the voice and the impact of the outside world could not penetrate. Often he forgot to bring the right books to class; often he stared at his masters without hearing their questions, remarks, or complaints. Sometimes, to conceal his isolation, he broke unexpectedly into a torrent of words which seemed incomprehensible or arrogant. He was relieved that mathematics had loosened his ties with Louis-le-Grand. But also loosened were the ties which bound him to his father, mother, brother, and sister, whose images grew more shadowy. The world of his thoughts began to destroy the world of flesh and blood.

With a perverse pleasure, he guarded the secret of his passion as though it would have been a betrayal to reveal it and a sacrilege to talk about it. He entered this new path alone, without friends, without encouragement, without understanding from anyone. Mathematics seemed to him an experience too great, too intimate, too personal to be shared. Only to himself he repeated proudly in his thought, "I *am* a mathematician."

When M. Vernier examined Evariste for the first time in mathematics, there was a rare silence. To his classmates who had peered at the titles of the strange books that Evariste read, this was the moment when a student might confound a boring teacher. To others, offended by his curt or arrogant replies, this was the moment when Evariste might receive a well-deserved humiliation. The silence puzzled and confused good M. Vernier. Evariste felt sick with disgust that he should have to perform before the class and answer utterly idiotic questions.

M. Vernier's manner was very friendly as he issued his first direction.

"Show how to divide an angle into two equal parts."

Galois felt the insult of this childishly trivial question. Red with shame, he drew an angle, then with a wooden circle he quickly sketched the arcs, lettered the diagram, and without saying a word, wrote:

$$ACE=BCE$$

"Very well done."

Then M. Vernier turned to the students.

"There are many of you who have been in this class over half a year longer than Galois and could not answer my question half as well."

Evariste's suffering expression deepened at these words.

M. Vernier asked, "Can you explain *why* these angles are equal?"

He emphasized the word "why" by raising the index finger of his right hand to the level of his nose.

Galois did not answer.

Patiently and kindly M. Vernier explained, "In geometry you must always show *why* something is true. You must always have a method, a good method of proving everything. Try to explain now *why* these angles are equal."

The friendly voice implied that he wouldn't even mind if Galois were unable to answer this question, that he was satisfied with what his pupil had done, and it would be quite all right if Galois only started to explain: then the master would be glad to give him a helping hand.

M. Vernier repeated, "Why are they equal?"

The class waited in suspense for Galois' answer. It came only after a long pause.

"Isn't it obvious?"

The whole class burst out laughing. Someone began to applaud. Someone shouted, "Geometry is obvious to Galois." Someone else cried, "Galois is obviously a genius!"

"Silence; silence." M. Vernier tried to calm his class. "You are very unkind to your fellow student. There is nothing to laugh about. Instead of helping him you make fun of your comrade."

Galois felt sorry for M. Vernier. He was a kind teacher, defending his pupil and not seeing, poor man, that the laughs were also directed against the master himself.

Evariste turned toward the blackboard, completed the drawing to two triangles, wrote down that they were equal, even indicated why, and deduced that the two angles were also equal.

M. Vernier looked at the blackboard with great satisfaction.

"Much better! Much better! Indeed this is very good. Try to work with more method. Only a little more method and you will be one of the best students in the class. But remember: be attentive and work systematically."

The school year ended. In the mathematics competition Evariste won second prize. M. Vernier was delighted. If Galois had only written more neatly, if he had only explained more fully, he might even have won first prize.

"A little more method, more method," thought M. Vernier, "and in a year he may even take part in the general competition."

Evariste also won a second prize in Greek in the general competition. When he learned this, M. Laborie murmured to himself, "Of course I am proved right. It did him good to repeat the second class."

During the next school year in the rhetoric class, only a few months after he had learned for the first time what the word geometry meant, Galois experienced the joy and suffering of creation. His days were full of tension and his nights were sleepless. It was the night that brought new ideas, which he turned over and over in his head, wishing he were allowed to light a candle and to write them down. When he did so in the morning, he often saw that his reasoning was faulty, that he had been kept awake by a mirage of the truth he was seeking. He worked on mathematics during study hours, he worked on his problems during other lessons, he worked while eating, he worked during the very few hours designed for relaxation, he even managed to work while writing an essay in French, or while answering his masters. At the back of his mind he felt the constant presence of his problems, even when reciting Latin or translating Greek. Whatever he did besides mathematics he did mechanically and without thought. His eyes were shadowed by dark lines, and his sight seemed to be directed inward toward his brain rather than outward to the external world.

What did the masters understand of their pupil? These are their notes for the first term of the rhetoric class:

Behavior fairly good. Some thoughtlessness! A character all of whose traits I do not flatter myself I understand; but I see it dominated by conceit. I do not think that he is vicious. His abilities seem to me far beyond the average with regard both to literary studies and to mathematics. But up to now he has neglected much of his work in class. This is the reason why he did not rank well in the examinations. He seems to have decided —from now on—to give more time and attention to work in class; together we have planned a new time schedule. We shall see whether he will stick to his own decisions. He does not lack religious feelings. His health is good but delicate.

To these kind words M. Pierrot added his:

Works little in my subject and talks often. His ability, in which one is supposed to believe, but of which I have not yet witnessed a single proof, will lead him nowhere. There is no trace in his work of anything but queerness and negligence.

M. Desforges wrote:

Always busy with things he ought not to do. Gets worse every day.

And finally there is a note by kind M. Vernier:

Zeal and progress very distinct.

5 : 1828

In 1823, Niels Henrik Abel, a twenty-one-year-old Norwegian, won fame in his native town because he was supposed to have solved the algebraic equation of the fifth degree. Later Abel found that his proof was wrong and, like every great scientist, stuck persistently to his problem: can an equation of the fifth degree be solved by radicals? That is, can the solution be expressed by a finite number of rational operations and extractions of roots on the coefficients of such an equation? Abel found the answer to his question. He published it in 1826 in the first issue of the journal for pure and applied mathematics, edited by Crelle in Germany. The answer was that an equation of the fifth degree is not generally solvable by radicals.

Galois, in the seventeenth year of his life, thought he had made a great discovery in mathematics. He believed he had solved an important problem, that he had a proof that every equation of the fifth degree can be solved by radicals. Later, after examining and re-examining his proof, he found in one lucid moment that his reasoning was faulty, and what he thought was a discovery gained by months of hard, persistent work, collapsed and turned into a heap of meaningless signs. But he did not give up; he knew, as all great scientists always have known, that the first weak ray of light comes only after constant, persistent thought; that the problem must be pursued for days and nights, for months and years; that one must think and think, wait and wait, until after incessant effort the first spark of understanding opens the narrow path that leads to a solution.

After fruitless attempts to solve the equation of the fifth degree, Galois believed that such an equation is not solvable by radicals.

Slowly the great problem of algebra began to crystallize in his mind: to find the right criteria which, applied to an algebraic equation of an arbitrary degree, would force this equation to disclose clearly whether it can or cannot be conquered by radicals. He was sure that if such proper criteria could be applied to a general equation of the fifth or higher degree, this equation would answer, "no, you cannot solve me by radicals." If the same criteria were applied to an equation of the third or even fourth degree, the answer would be, "yes, you can solve me by radicals."

Thus Galois, a student of the rhetoric class at Louis-le-Grand, formulated one of the most difficult problems in mathematics, one of the greatest problems in algebra. Yet he could hardly have known how important this problem would prove to be; he could hardly have known that the revolutionary and powerful methods by which he would solve it would influence the development of mathematics a century later.

Faithfully every term, the professors recorded their notes. At the end of the second term, the master supervising Galois' studies wrote:

> His conduct is very bad, his character secretive. He tries to be original. His abilities are distinguished but he does not use them in the rhetoric class. He does absolutely nothing for his class. A passion for mathematics possesses him. I think it would be better for him if his parents would agree that he study only mathematics. He is wasting his time here and only torments his masters and constantly receives punishment. He does not lack religious feelings; his health seems poor.

M. Pierrot wrote:

> He has been doing some homework; otherwise talkative as usual.

M. Desforges wrote:

> Erratic, talkative. I believe that his ambition is to wear me out. He would be very bad for his classmates if he had any influence on them.

M. Vernier, the mathematics teacher, wrote:

> Intelligence and progress marked; but not enough method.

When the year in the rhetoric class ended, Evariste knew well what he would do next: he would enter the Polytechnical School.

Often during long nights, Evariste suppressed from his tired mind thoughts of permutations and their products, roots written in the form of continued fractions, by thoughts of his near future in which he saw himself wearing the uniform of a student at the Polytechnical School.

The Polytechnical School! The daughter of the Revolution and the pride of France! There he would be allowed to work all day on mathematics. Even more, he would be *obliged* to work all day on mathematics. He would meet men who would understand him: the greatest mathematicians of France—some of the greatest mathematicians in the world. He would listen to Cauchy's lectures. Cauchy would recognize the importance of the problems on which he, Evariste Galois, was working. He would meet Ampère and François Arago, admired by the students and loved by the people of France.

He would meet new comrades and make new friends. It was true that he had not made friends at Louis-le-Grand, but he would make them at the Polytechnical School. In a few months he would start a new life, his real life, in the school that educated not only scientists, employees of the state, and officers in the army, but also leaders of the people. He knew that to enter the Polytechnical School he had to pass an oral examination. What a pity that it was oral and not written. He must pass! He would have to reveal his knowledge to the examiner, perhaps even the problems on which he was working and the results he had achieved. The thought was disagreeable, even painful.

He remembered his last talk with his father, to whom he had revealed his love for mathematics and his plan to enter the Polytechnical School. His father understood him; no one else ever did. His teacher in mathematics had taught him for over a year now and had never suspected how much he could learn from his own student.

But his father had understood. His eyes had lit with pride when he said gaily, "My son will be a great mathematician. Evariste Galois, professor of the Polytechnical School, member of the Academy. Yes, Evariste, these are fine-sounding words. I like them."

Then he burst into laughter, but the laugh was short-lived and somewhat forced. It was not like old times. His father's eyes clouded quickly when he said, "I hope that in your life you will not find as many enemies as I have." Then he spoke very softly as though to himself. Evariste could hardly hear him. "This is still not the worst. Indifference is the worst."

He turned quickly toward his son.

"Here I am, spoiling your pleasure. M. Vernier wrote me that you are very good in mathematics. He did not sound as stupid as you make him. His advice is that you stay a year longer at Louis-le-Grand and take the mathematics special as everyone does who wants to enter the Polytechnical School. What do you think?"

Evariste was angry with M. Vernier for meddling in his affairs, and even disappointed with his father who was ready to leave him at Louis-le-Grand for another year.

He was surprised how cold his voice sounded when he asked, "Don't you believe me, that I know enough to pass the silly examination?"

"M. Vernier wrote me that you may know too much to pass the examination; that you know the important things, but that you may not know the unimportant details which the examiners always ask. His advice is definitely for you to stay a year longer at Louis-le-Grand."

"M. Vernier is old and stupid." Yet he wished now that he had not said these words.

6 : 1828

It was the hour which the students at Louis-le-Grand spent writing letters to their parents, friends, and relatives. Everything was rigidly regulated at Louis-le-Grand; even the love of children for their parents.

Evariste wrote:

Dearest Father:

A week ago I wrote you a distressed letter which must have upset you. But your calm and kind reply helped me very much. Now I feel less unhappy and more composed. It was a difficult week! When I failed in the entrance examination, I felt at the end of all hope, the end of life itself. Then I repeated your words. It was good of you to say that you were afraid of this not because you did not believe in me, but just because you *did* believe in me.

I now understand that you thought this might happen when you advised me to remain a year longer at Louis-le-Grand and to take the special course in mathematics. So here I am at Louis-le-Grand for another year! It seems to be my fate to spend my life in this prison building that I know so well and detest so much.

I never thought that I could feel such contempt for any of the men to whom I looked up some months ago as I now feel for M. Lefebvre, my examiner. He is a poor scholar and his face looks like a skull over which a wrinkled yellow skin is drawn. He looked most repulsive and inhuman to me from the first moment I set eyes on him. This examiner of my dream school hissed his silly questions; I saw from his tone and look that a student is mud under his feet. I am sure that he is a Jesuit. What this yellow skull expected was a quick recitation of formulas without understanding. He wanted to have everything explained the same way as in the silly textbooks. He considers it a crime to have original ideas and methods of presentation.

When my turn came, he looked at me with his small eyes, then he closed them still further so as to see as little as possible of me. Then he asked the first question.

"Why did you come to the examination without the special mathematics course?"

I answered, "I studied by myself."

"Oh."

You ought to have heard that "Oh!" Then he asked me how to solve an equation of the second degree. He dared to ask me, who knows more about algebraic equations than all the professors at the Polytecnical School taken together, this insulting question. And besides, his question was wrongly formulated. When I said, in answer, that the question was badly formulated, the yellow skull wrinkled its skin in what was intended to be an ironical smile. Then he waved away my remark, saying that he had no time for discussion and that it was not he who was to be examined. Then he asked me the most childish questions. I felt a contraction in my throat and could not utter a sound. The skull then turned to me.

"I see that you studied by yourself; but you did not study enough. You had better try again next year."

Dear Father! I shall listen to you and I shall try to learn the silly little tricks and to answer next year in the language they expect to hear. And I hope to be more successful then.

Let me now leave this unpleasant subject.

Dear Father! You seemed depressed when I last saw you. I am grateful that you told me something of what is worrying you. It only confirmed some of the suspicions that I had before. But the men who have started a campaign against you, the lowest and most detestable campaign of slander, those men shall not succeed! They will never succeed in smearing your honorable name! The people of Bourg-la-Reine know their good mayor and they will not listen to the calumnies of the parish priest. The Jesuits may be strong, but not strong enough to turn away the people who love you.

Galois paused here and read his last words. They did not sound right. They would not bring the relief his father needed. He wrote:

My dear Father! How much I would like to help you with my love! To repay you for your love, friendship, and understanding! Instead I am only adding to your sorrow by my own misfortunes. But I believe with you that times will change. A storm will come that will clear the air here in Paris, in Bourg-la-Reine, and in the whole of France. Let us hope that it will come soon.

The bell rang. It was time to finish the letter. Evariste wrote quickly.

I send you my love. Please explain my failure as well as you can to Mother. My love to all of you,

Evariste

He then went to the small room of the special mathematics class where, among some twenty others, he waited for the first lecture of the new professor of mathematics.

When M. Richard entered, he did it without dramatic effect. After closing the door, he smiled dreamily and seemed to hesitate. Then he went to the rostrum, turned his broad, slightly curved back to the class, took a piece of chalk, broke it in two and turned around. He looked absent-mindedly at his students, who examined the big man, his thinning hair, square head, and friendly eyes blinking through thick lenses. When he began to speak, he did it very calmly, without oratory. Some of his listeners wondered how this man, who talked as casually as to friends in a small drawing room, could be known as the best teacher at Louis-le-Grand. Yet everyone listened.

"In this course, my young friends, our purpose will be to broaden our knowledge of mathematics. We shall try not only to broaden but also to deepen our knowledge. We shall try to achieve it by starting from the beginning once again. We shall quickly review the material you have learned, but from a more advanced and modern point of view. Such a quick review will allow us to see the essentials, the most fundamental theorems upon which others are based. There is a great danger of seeing the trees and not the woods in mathematics, of seeing small theorems and forgetting the structure of the subject in which these theorems appear and by which they are linked together."

Evariste was prepared to spend the hour of M. Richard's lectures on his own work, but now he listened.

"Let us devote our attention to geometry. When you learned it for the first time, you must have had the impression that geometry, complete and finished, sprang suddenly from the brain of one man, perhaps even in book form. But geometry, like any other branch of mathematics, is the result of the work of generations of men. It is connected mostly with the name of Euclid, who lived about 300 B.C. But geometry started a long time before Euclid. And perhaps you will ask me: When did it end? It did not end and I do not believe that it ever will."

Here M. Richard outlined the story of geometry: how the Egyptians began it as a practical science of measurement, and what role the Greeks played in its development.

All this was new to Galois. He would not admit that it is important for the understanding of mathematics to know its history; but he had to admit that he found everything that M. Richard said interesting and he liked the way it was said: and—this was the greatest compliment Galois could pay—he listened.

"One of the gravest dangers in teaching is to convey the impression that mathematics is like a sealed book, like a finished structure given to us by past ages, to which nothing can be added and in which nothing can be changed. Mathematics is a living organism. And especially in modern times, in our nineteenth century, it is living vigorously. Even elementary geometry may become a source of new and very important discoveries.

"You may think, my friends, that creation and doubt can come only after you have mastered the subject. You may think that only when you have absorbed all the knowledge in a branch of mathematics, only then, perhaps, your own ideas may come into play. This may be true as a rule, but does not need to be. Again geometry forms a very good example. Here we see our doubts and troubles starting right at the beginning. We shall understand this better if we consider in a few words the history of Euclid's postulates, or, as we call them now, axioms."

Here M. Richard enumerated the five Euclidian postulates, analyzing each of them in turn until he came to the fifth axiom.

"The story of the fifth axiom leads us straight into modern times. This axiom never seemed as self-evident as the four others. Many attempts have been made to replace it by some other axiom which might appear more self-evident. Whether we can prove the fifth axiom or whether we must assume it, and what is the best form in which to do either, all this and other problems are still open; and the future may bring new and unexpected discoveries."

Evariste thought about the difference between M. Richard and M.

Vernier. He had to admit—very reluctantly—that he might even learn something from his new teacher.

"M. Richard," thought Galois, "is not a great mathematician himself, but he likes mathematics and has breathed its air with love and understanding. Even if he has not done much creative work himself, he sees its beauty and he knows how to make others see it." Evariste decided that M. Richard was a man worth knowing, a man to whom he, Evariste Galois, might even reveal his powers.

M. Richard dictated the weekly collection of problems. They were regarded as difficult by the majority, requiring many hours of work, and even the good pupils seldom managed to solve all of them.

The students copied in their notebooks: Problem I: Find the two diagonals x and y of a quadrilateral inscribed in a circle, in terms of its four sides a, b, c, d. Then they wrote down the second and third problems. Evariste only listened, and by the time the dictation was over he had the detailed solution of each clearly before his eyes. M. Richard then began to lecture.

Evariste tore a piece of paper from his notebook, wrote on the top "Galois" then below "Problems." He formulated the first and wrote its solution by means of equations and explanations forming the concise links between these equations. Without crossing out or correcting one single word he arrived at the result in the simplest possible manner, writing down explicitly the values for xy and $\frac{x}{y}$. Then on the next page he wrote equally carefully the rigorous solution of the remaining two problems, providing each with a neat drawing. All this took him fifteen minutes, after which he only half-listened to M. Richard's lecture, more occupied with collecting his courage for the end of the hour.

As M. Richard was leaving the classroom he heard, "Excuse me, M. Richard."

"Yes?"

The teacher saw a student, thin and small for his age, with a blushing triangular face, staring at the floor and holding out a sheet of paper.

M. Richard put his arm on Evariste's shoulder and asked, "What is it?"

Without looking up, Evariste gave the sheet of paper to M. Richard.

"Here is the solution."

M. Richard looked at the first page, quickly read it through and saw the problem solved in a style worthy of the best textbook. He turned the page, looked at it, then at the student, then again at the paper, and then again at Galois. He turned back to the first page and read the signature aloud.

"Galois. What is your first name?"

"Evariste."

"I see."

He looked at Evariste for a long time without saying a word. Evariste felt ashamed and regretted what he had done. Had he made a fool of himself? Would M. Richard now smile ironically as the yellow skull had done?

M. Richard said, "What about coming to my room today after dinner, so that we can have a long chat? I shall ask your master not to crucify you if you return a little later to your dormitory. Agreed?"

"Yes, sir."

"Good."

Galois was burning with excitement. As he turned away he heard one of the students whistling and saying to his neighbor, "Imagine! Our genius tries to make friends."

And he heard, too, the neighbor's reply: "I'm afraid it will kill him."

Like most of the professors, M. Richard lived at Louis-le-Grand. When Evariste entered his study, M. Richard indicated a chair opposite him, looked at his guest for a while and then, as he filled his pipe, said, "I wish you would tell me something about yourself. What are you working on?"

The secret of M. Richard's success with students was very simple and consisted in one guiding principle: to treat them as equals.

Evariste was astonished that he did not need to convince M. Richard that he was a mathematician. In some strange way M. Richard seemed to know it. For the first time at Louis-le-Grand, Evariste felt shy and humble.

"I am working on algebraic equations. A year ago I thought that an equation of the fifth degree is solvable by radicals in the way equations of the third and fourth degrees are. Now I believe that the general equation of the fifth degree is not solvable by radicals."

Galois stopped. M. Richard looked with astonishment at the student opposite him but merely said, "Hm! Very interesting! Very interesting."

"The problem on which I am working is really much more general. I am looking for necessary and sufficient conditions for an algebraic equation to be solvable by radicals. I mean for an algebraic equation of arbitrary degree. I believe, I am really quite sure, that such criteria must exist." Then he added confidentially, "I believe, sir, I have recently made some progress toward the solution of this problem."

He was anxious to explain his result in detail, but was only slightly disappointed when M. Richard looked at him in silence for a long time, and then said, "This is an ambitious plan." He took a puff at his pipe and repeated, "This is an extremely ambitious plan. You know, my young friend, that if you solve this problem, you will be in the ranks of the best mathematicians of our generation. I wish you luck and success with all my heart. By the way, how old are you?"

"I was born October 25, 1811."

"Seventeen years ago. Seventeen years old. I am almost exactly twice your age. Tell me more about yourself. How did you manage to grow so old without solving the fundamental problem of algebra?"

He laughed broadly at his joke and the laugh affected Evariste.

"When did you become interested in mathematics?"

Galois now talked louder and more freely. He told M. Richard about Legendre, about M. Vernier, about the examination at the Polytechnical School, and even about his home and his father.

It was late in the evening when M. Richard said to Galois, "You can do much for me, my friend. You can help me to arouse interest in mathematics in your class. You see, this is the problem: at most of my lessons you will be extremely bored. You already know—with the exception of some trivial and unessential details—everything that I intend to say and, of course, much more. As a matter of fact, I am not ashamed to admit that in some branches of mathematics you may know much more than I do. Now the question is how to save you from boredom. And boredom is a contagious disease. You may unwillingly spread this disease through the class, and that would be bad."

Evariste interrupted:

"Oh! I shall never be bored at your lessons, sir."

"Of course you think so at this moment. But you may feel differently in a few months. Yet I believe there is a way out. What you have learned, you have learned by yourself—not so much in the school, but perhaps in spite of school. Try to think about the lessons, not from your own personal point of view. Think that their purpose is to create an interest in mathematics, not only to teach it, but to make it living and exciting. If you know the subject of my lectures perfectly, then ask yourself whether my presentation is sufficiently clear; and if you have any critical remarks to make, please make them."

"Oh! I wouldn't dare to do that."

"But that is just what I want you to do. Discussion increases interest, so does the atmosphere of doubt and argument; it leads to clarification and deeper understanding. In this way the lessons will become for all of us experiences to which one looks forward with anticipation, and remembers pleasantly when they are gone. By your attitude, you can help me to create the right atmosphere."

"I shall be very happy, M. Richard, to do whatever you ask."

He wanted to say, "You, sir, are the first at Louis-le-Grand to be kind and understanding to me." But these words remained unsaid.

7 : 1829

Evariste Galois made his scientific debut while still a student at Louis-le-Grand. His first paper was published in the *Annales de mathématiques de M. Gergonne,* and its title was: *Démonstration d'un théorème sur les fractions continues périodiques.*

It was greeted by silence. Evariste told no one about it, and no one seemed to be bothered by its appearance. True, it was not a very important paper. It was not in this paper that Evariste formulated his results on the solvability of algebraic equations. This he did in a manuscript that he sent to the French Academy, a manuscript that contained some of the greatest mathematical ideas of the century. For the first time at Louis-le-Grand, he had a feeling of relaxation and happiness. Yes, he knew that he had formulated the paper concisely. But any great mathematician ought to see that the manuscript must be read slowly and studied carefully. The paper would probably be sent to M. Cauchy. Evariste was sure that the great master would recognize the importance of the results, and of the methods by which they were achieved; he would see that this paper opened a path that would lead to still greater discoveries. Soon the whole world would know what, until now, only he knew: that he, a student of Louis-le-Grand, who had not passed the entrance examination to the Polytechnical School, was a great mathematician. Even M. Richard, even his father would be astonished. Soon he would become a famous mathematician, famous not only in France, but all over the world, everywhere where mathematics was taught and studied.

He spent much time daydreaming, imagining how Cauchy would receive his manuscript. His favorite reverie always started with the arrival of his manuscript at the home of the mathematician.

"M. Cauchy will first say to himself, 'Ridiculous! A student of a college sending a manuscript to the French Academy!'

"But M. Cauchy is a great mathematician and he knows his duty as a member of the Academy. He will begin to read the manuscript. His interest will increase with every word; his astonishment with every page. He will see the importance of the distinction between a primitive and a nonprimitive equation. Good that M. Gauss is quoted in the first few sentences. At least M. Cauchy will be sure that the author knows the literature and does not merely rediscover well-known results.

"M. Cauchy will recognize that a new path has been opened through the unknown. He will become more and more excited. He will immediately write a letter to M. Gauss. Then he will write to the Academy. No, he will not write to the Academy or to M. Gauss. This will come later. His first impulse will be to meet Galois, to embrace him, to congratulate him, to ask him to his home, to ask him on what he is working now and what are his plans. But even this will come later. First he will have to find Galois at Louis-le-Grand. To do this he will have to go to the director's study. He will see M. Laborie and say, 'I am M. Cauchy.'

"Then M. Laborie will bow very deeply. He will ask humbly to what he owes the great honor of M. Cauchy's visit to Louis-le-Grand.

"And M. Cauchy will reply, 'Do you know that you have a genius in your school? He has solved a problem on which I have been working for a long time and could not solve. May I see him? His name is Galois.'

"And M. Laborie will reply, 'Oh—Galois? Of course, M. Cauchy. He is the pride of the school. We love and admire him. As a matter of fact, we love him so much that we have kept him for two years in the second class.' "

Then Galois' daydreaming turned toward the Polytechnical School. Next year he would again compete and take the entrance examination. Perhaps the same yellow skull would examine him. But this time it would be different. He would look at Galois with amazement and say, "Are you *the* Evariste Galois?"

"What do you mean by that?"

"I mean *the* Galois who wrote the famous paper on the solvability of algebraic equations?"

"Yes. I am the same man! I am the same man you turned down a year ago at the entrance examination."

"How is it possible? Oh, M. Galois! You must forgive my stupidity. I shall be the laughingstock of the country if it is found out. I failed Galois, one of the greatest mathematicians of our time! And you are only seventeen! What will happen if you turn out to be the greatest mathematician of all times? Then I shall become famous as the man who did not pass Galois."

"Precisely, this will be my revenge."

Why did he think of all these stupid, childish scenes instead of his father? Why did he not think about his father first? He would say to his father, "Father, do you know? I am famous. I am a famous mathematician."

And his father would answer with a smile, "I always knew it would happen. I always believed in you."

The academician, M. Cauchy, opened a boiled egg absent-mindedly and at the same time looked through the manuscript of one of the seven hundred and eighty-nine papers that he wrote during his life. The day was too short for M. Cauchy to put on paper all the ideas that burned in his brain, to prove all his theorems, to prepare all his lectures, and to perform all his religious duties. One must work and pray in life, but M. Cauchy worked too hard and prayed too long.

M. Cauchy's wife was plain, silent, and as pious as her husband. She came into his study, put the mail on his desk, and went away. M. Cauchy did not have time to raise his eyes or smile at his wife. He looked at the manuscript for corrections and opened his mail mechanically, Another manuscript from the Academy! He glanced at the signature and at the words below it: "Student of Louis-le-Grand."

"Soon they will send me papers written by babies in diapers. Why do they send me all the crazy papers about trisecting an angle or the solution of some great problem by men who never did any solid work before. Don't they know that my time is too precious for me to try to cure these potty brains?"

He threw the manuscript into a wastepaper basket.

"It's good I didn't notice the name. Tomorrow I'll surely forget the whole incident. When the secretary asks me what happened to the paper of this potty mathematician, I can say in all honesty that I don't have the slightest idea and I don't recall any such name. And I won't be lying."

But M. Cauchy felt disturbed. He did remember now that sometime ago he had thrown away another manuscript. It was a paper written by a foreigner, and M. Cauchy did not like foreigners. But unfortunately this foreign name stuck in his mind. Why was he so stupid as to read it? A curious name, a Biblical name, very difficult to forget. Yes, it was Abel's manuscript. Why didn't they send it to Cain? He tried to laugh at his joke, but he did not find it very amusing. He turned to his own paper, brushing aside the thought of Abel, Cain, and the student at Louis-le-Grand.

8 : 1828–1829

Louis XVIII once said about his brother, Comte d'Artois, "He conspired against Louis XVI, he conspires against me, he will conspire against himself." And conspire he did! He conspired against himself when he conspired against his Prime Minister, Martignac, calling him "a beautiful organ of speech" and scheming for his downfall. He could not forgive Martignac for trying to come to terms with the opposition of the moderate liberals and for not placing the crown above the chamber; for seeing clearly that the power of the bourgeoisie increased with the real and imaginary growth of the Jesuit bogey. Thus the King forced his Prime Minister to resign and appointed the last Prime Minister of the last Bourbon king: Prince Jules de Polignac.

When we look at Polignac's picture, we see a striking face. The head is long, thin, and supple; the features are aristocratic and sharp, the nose long and distinct. We can almost feel that to indicate any object in his vicinity he would, in a refined manner, use his little finger. His hair falls on a forehead that appears disproportionately small; the eyes seem to look through the real world straight into the faces of imaginary angels. On the lapels of his elegant coat small lilies are embroidered. Around his long neck he wears a white silk cravat tied like a muffler and framed by a silver waistcoat with longitudinal black stripes. There is a striking similarity between Polignac and Charles, whose illegitimate son he was supposed to be.

Prince Jules de Polignac was the counterrevolution incarnate. Only the ultra ultras and members of the congregation rejoiced over the King's choice.

The new Prime Minister was the son of the scheming Duchesse de Polignac, a favorite of the guillotined queen, Marie Antoinette. For forty-nine years he had carried with great pride the burden of his family's extreme unpopularity. When asked how he could administer France without a majority in the chamber, the Prince said that he wouldn't know what to do with it if he had one. He refused to listen to any advice if not given by the King or the Virgin Mary, with whom he claimed to converse in his dreams.

France expected shattering events. Yet for a few months nothing happened. France was like a great theater where an impatient public crowds in to see a play and the curtain refuses to rise. Perhaps the one new thing that did happen was that a new word was added to the Parisians' dictionary.

A carter commanded his horse to move, but the horse was obstinate and refused; even whipping did not help. The exasperated carter shouted, "Go on, you Polignac." The horse moved. Since then, obstinate and stupid horses were called "polignacs" by the Parisians.

The obstinate and stupid Prime Minister drove the carriage of state with the King in it, while the revolution waited around the next corner.

IV

PERSECUTION

1 : July 2, 1829

When Evariste opened the door to the director's study, M. Laborie rose at once. He placed his hand on Galois' arm and asked him to sit down. His lips were tightly closed, but his face seemed softer than usual, as though he wore a mask of sympathy and pity. Without speaking he returned to the desk and picked up an envelope which he fingered as he sat down opposite Evariste.

"I have some sad news for you, very sad news. A letter from your father arrived with a short note to me. You must be prepared for sad news and you must meet it with courage. We are all in the hands of God and it is at such times that we should turn to our Saviour for consolation and pray for His blessing. I want you to know, Galois, that you have our deepest sympathy, mine and all your teachers'. Go into the conference room and read your letter undisturbed."

Galois went into the adjoining room, tore open the envelope with unsteady fingers and began to read:

My dearest Son,

This is the last letter you will ever receive from me. When you read these words, I shall not be among the living. I do not want you to despair or to mourn. Try to make your life normal and full as soon as you can. I know that it will be difficult for you to forget your father who was also your good friend. But I want you to spend as little time mourning and brooding as you possibly can.

I am leaving you an income which should provide for your studies. The rest of my family also will be modestly, but sufficiently, provided for.

One often thinks of suicide as a cowardly act, an escape to which man has no right. Perhaps this is true. But the burden of life has become unbearable to me. Only death can bring peace and end my sufferings. My dear Evariste, when you read this letter think that I am now beyond all suffering, that no one can hurt me now, that through my death I shall protect you better than I could by living.

I shall try to tell you as well as I can why I have decided to take a step from which there is no return. It is perhaps the only human deed that can never be undone.

You know, my son, that for seventeen years I have been the mayor of our town—before, during, and after the hundred days of Napoleon. After Waterloo, the enemies of freedom tried to remove me but they failed. Everyone knew my convictions and what I thought about the Bourbons and Jesuits. But in spite of my opinions, I remained the mayor because I had what no one else had at Bourg-la-Reine: authority. Now, my dearest son, when I look back on the old days, I see in them an honorable fight with my adversaries and also achievement, happiness, and honor. Whoever fought me then did so openly and I defended myself openly.

You remember, my son, how often men and women of our town came to their mayor for advice, and you saw the respect and confidence that they always showed me. It was on their respect—and not fear—that my authority was based. There were times when some of the citizens wavered. They were confused because they heard different things from the priest and from me. Some of them could not make up their minds and switched from one side to the other. Some of them were always against me, but the best citizens of our town faithfully remained with me.

You must have noticed the change in me that began two years ago when the new parish priest came to Bourg-la-Reine. Perhaps it was my fault that I never talked to you about it, but I found it too difficult. Suddenly I found that I breathed poisonous air, and I feared that I should have to breathe this poisonous air for the rest of my life. I felt that only the tomb and the earth of my town could protect me by covering my dead body.

I am sure, my son, that the parish priest and the men who sent him here knew that they could not undermine my authority in an honorable fight. They changed their methods. They no longer called me a Republican, nor a Bonapartist, nor even a Liberal. These names disappeared from their vocabulary. On the surface it looked as though they had ceased to fight. I was no longer a dangerous adversary whom one fears. They pictured me as ridiculous, a crackpot, a crazy man, a pathetic figure who belongs in a lunatic asylum. Some people began to greet me with poorly concealed smiles. Others, who have always been against me, laughed in my face, singing fabricated couplets about Bourg-la-Reine, a town at which the whole country laughs because it has a crackpot for a mayor.

But perhaps the worst was the expression of my former friends. There was pity in their looks. Pity! I became afraid to talk to you for fear I might see pity in your face. If, following an old habit, I quoted a sentence from Seneca or Voltaire, my friends lowered their eyes and blushed. The children of the town were the worst. They were taught to stand in front of my house and sing about a "mayor who was crazy all his life and so were his children and so was his wife." When I did not react, I was laughed at. When I tried to use persuasion, I was laughed at. When I reacted with outbursts of anger, I was doubly laughed at.

You remember, dear Evariste, how in the good old days we amused ourselves by writing couplets about the people of our town and the events of our times. These rhymes were sometimes spiteful, sometimes witty, and sometimes silly. Some of them circulated through the town and people liked them. In the last two years, the most vulgar and indecent couplets have been circulated and attributed to me. Even some of my friends have acually believed that I wrote these filthy rhymes. Those who were decent enough to ask me, may have been convinced by my denials. I say "may have been," because I do not know for certain, and I shall never know now.

I am sure you are surprised at the devilish simplicity of their plan. So was I. I cannot understand now why it took them so long to hit upon it. In Bourg-la-Reine our family lived differently from most of the others. We had our books; we had our convictions. We talked and lived in a way which others could either respect or laugh at. They chose to respect me for fifteen years, and to laugh at me for the last two.

I thought of leaving the town and going to Paris. Indeed, you know that recently I often came to Paris where I leased a small apartment. It is here, so near you, that I write this letter. But the laughs, the shrieks, the songs followed me. And to go away for good would mean to admit defeat. There is only one way to awaken the conscience of those who caused my misery: to take my own life so that they will know why I did it. By taking this last step I may restore the respect which they have felt for me and my family. No one will then dare to laugh at your mother or you.

I shall die by suffocation. I shall die because there is not enough pure air for me to breathe. This poisonous air by which I shall die here in Paris was fabricated by men at Bourg-la-Reine. This must be known and understood.

It is difficult for me to say good-by, dear son. You are my eldest son and I have always been proud of you. Some day you will become great and famous. I know that the day will come. But I also know that suffering, struggle, and disappointment lie ahead of you.

What happened to me is not accidental. You understand, my son, that it is not the parish priest and not the stupidity or viciousness of some people which force me into the grave. These are—as you know too well—only external signs of something much broader and deeper.

You will be a mathematician. But even mathematics, the most noble and abstract of all the sciences, has its crown in the air but its roots deep in the earth on which we live. Even mathematics will not allow you to escape from your sufferings and those of your fellowmen. Fight, my dear son, more courageously and more gallantly than I did; and may you hear the bells of freedom ringing in your lifetime.

M. Laborie entered the room. He came near Galois, stroked his head with a fatherly gesture and said, "Wouldn't you like to see M. Richard? He has told me how much he thinks of you. It may do you good to see him."

With an effort Evariste spoke through his sobs.

"No! I don't want to see anyone. I want to see my father."

"Be calm! I know how you feel. If you wish, you can stay out of school for a week. You can go immediately if you want to. I shall tell your master to help you."

After Galois had left the room, the director murmured to himself, "These are the results of atheism. No believing Catholic would commit suicide. How can our school restore religion if the family destroys it? A father making his own son so unhappy! It is the curse of atheism that haunts our time. I am sorry for the poor boy. He is the victim!"

Then M. Laborie sat down at the desk and continued his work.

2: July 5, 1829

The funeral procession moved from the mayor's house toward the church. The hearse with its cross and angels carved in black wood was flanked on each side by three pallbearers and driven by two men in black uniforms.

Mme. Galois, her daughter, and Evariste walked behind the coffin. Evariste's mother held erect her cold, strong face. Behind them came Mme. Galois' sister, and Alfred, Evariste's younger brother, other members of the family, and the citizens of Bourg-la-Reine.

The citizens whispered: "Yes! It was the parish priest." "It all started when he came." "Even if the mayor was a little crazy, he was a decent man and he was our mayor." "Will the priest dare to come?" "Will the priest dare not to come?"

Those citizens who had hated the mayor now turned their hatred against the priest. Why had he not foreseen that the mayor had one trump card to play: his own life? Those citizens who had loved the mayor now doubly hated the parish priest.

The procession neared the church. The spectators looked with hungry eyes to see whether the parish priest was among the group of men and boys in black and white standing before the church to receive the body. No, he was not there. "He is a coward," whispered those who would have whispered "How does he dare?" if they had seen him. Some saw with relief, some with anger, the curate standing in surplice and with the ritual in his hand. It was he who would pray for the mayor instead of the parish priest, and it was he who would sprinkle the dead mayor's body with holy water. An altar boy who stood between the curate and two priests from neighboring parishes carried the holy water font with a hyssop. Acolytes with candles flanked the group and an altar boy held a cross before it.

The hearse stopped and the pallbearers lifted the coffin. The dead and the living filed into the church, led by the altar boy's cross and the curate's *Miserere mei, Deus.*

Many of the citizens remained outside. They showed their disapproval by not entering the church which had come between them and the mayor they had loved. Again they repeated their accusations. But by now words were not enough; they must demonstrate to the clergy who had hated the mayor, how much they, the citizens, had loved him. They moved to the front of the hearse and stood there, watching the carved doors of the church.

Now they saw the cross, the clergy, the coffin, and all those who had entered the church before. Reluctantly, with ill-concealed hostility, they made room for the curate, priests, and altar boys. The curate prayed and the funeral procession moved toward the cemetery.

Suddenly the men who preceded the hearse stopped. By doing this they forced the altar boy, who carried the cross, to stop. When he stopped, the clergy stopped, and the hearse with the coffin stopped. The whole procession stopped. Then some of the men who stood before the hearse rushed to the coffin. The suddenness of their movements seemed out of place, indecent at the funeral. Ignoring the pallbearers, they took the coffin. Defiantly one of them announced, "We shall honor the mayor by carrying the coffin."

The curate looked back. He waited calmly, showing neither anger nor disapproval. The demonstration which was intended both to honor the mayor and anger the clergy succeeded only in honoring the mayor. The curate's calm affected the others. No one protested. Soon an orderly procession moved ahead, the coffin now carried by men and not drawn by horses.

They came to the small church at the cemetery. Here stood the parish priest: calm, proud, erect, in surplice, with a biretta on his head and the ritual in his hand. Everyone understood that it was the parish priest who would say the last prayers and who would bless the grave.

A voice was heard: "We don't want the parish priest."

Voices responded: "Away with the parish priest. Down with the Jesuits."

The priest moved calmly toward the coffin and took his place before it as though he had heard nothing. After a few repetitions, the cries died down, but the tension grew. The men who carried the coffin looked with hatred at the stiff back of the priest that they saw before them. Only the weight of the mayor's body kept their lips sealed. But soon they laid down the coffin beside the spot where it would stay forever. Now their hands were free. They stepped aside and let the members of the family approach the coffin. The parish priest and the clergy stood on one side; Evariste, his mother, and his sister on the other side, with the length of the mayor's body between them. All those who stood near the clergy moved away. Some of them went home; they did not wish to witness the events that might come. Others moved to the side where the family stood, to increase their distance from the parish priest and to show that they disapproved of his presence.

The priest began to pray: *"Deus, cujus miseratione animae fidelium. . . ."*

A loud voice interrupted, "Murderer."

Someone repeated, "Murderer."

Then other voices: "You murdered our mayor."

The priest looked up from the prayer book, straight into the faces of those who stood opposite him. Then he raised his eyes to heaven.

"Oh God, forgive them, for they know not what they do."

Then his eyes moved firmly from one face to another.

"Here at the coffin of the mayor, we stand united by pity and forgiveness. The ways of our Father in Heaven seem strange and incomprehensible. We must accept His will with humility for we cannot understand the wisdom of His judgments. I was sent to you by God's representative on this earth. Who of you has the tragic courage to say that the Church is—or that I am—responsible for the mourning of the mayor's wife and of his children? Do we not show good will, pity, and forgiveness by coming to the grave that I am now willing to bless? Does not our religion forbid us to be masters of our own lives? The unfortunate mayor took his own life because his poor soul and mind were tormented by unhappiness from which faith alone can save us. But we came here with you to bury the mayor on consecrated ground because it is the duty of the humble servants of Christ to have pity and to bring consolation to those whom God has commanded to bear the burden of life. It is for their sakes that I came here to pray for the peace of the deceased's soul. And may God Almighty have mercy on all of you who have raised your voices against me.

"Let those who dared to throw into my face that most horrible accusation come forward; let them show their faces to me and to the rest of you. Let them then dare to repeat the accusation if this is what they believe."

Evariste waited for the men who had cried "murderer" to throw the word again into the priest's face. He felt vaguely that he had experienced a similar scene before at Louis-le-Grand, only now its sorrow and tension was a thousand times magnified, made a thousand times more morbid by his father's dead body closed in the coffin which lay between him and the priest.

No one repeated the accusation. Evariste closed his fists, digging his nails into his palms, but he could not create a pain strong enough to diminish his flaming hatred. He felt the pressure of his mother's arm, and looked at her face. Her calm was gone; her face was lined by suspense, and fear was in her eyes.

The priest raised his voice once more. In it Evariste heard both triumph and mockery.

"Is there anyone among you who believes that the Church and I can in any degree whatever be held responsible for the tragedy that has happened?"

Evariste freed his arm from the grip of his mother's hand. He moved one step forward and felt the coffin touching his legs. He looked into the priest's eyes and said, "I do."

The words removed the spell. Now angry shouts flew from all sides.

"Murderer! Murderer!"

Someone threw a stone at the priest. The clergy moved back, without panic, steadily increasing the distance between them and the coffin. But the parish priest stood immobile, his eyes raised to heaven. The stones flew more densely and the word "murderer" grew louder. Some of the stones fell on the coffin. Then one hit the priest's forehead. He fell and blood ran from his face. The curate and an altar boy knelt, trying to raise him. Stones were still flying.

"Stop it! Stop it!"

It was Evariste's mother. Her face became distorted in an hysterical grimace.

"Stop it. For God's sake, stop it!"

Evariste felt that his legs could no longer support the heavy weight of his thin body. He fell down, embraced the coffin, and cried in a voice that became more hysterical with every word that wracked his body, "Oh, Father; my dearest Father. Take me with you. I don't want to live. No! No, it isn't true. I do want to live. I will live as you wanted me to live. You will always be with me, whether I am dead or alive. Oh, dear Father. I shall think of you always, my whole life, to the last moment of my life. I swear to you that I shall never forget what you said to me and what you taught me. But I hate, Father. I must hate. Do you hear me? You must forgive me. I hate all those who fought you. I must hate. I must hate!"

The words grew more and more inarticulate until they ended in a crescendo of sobs and cries in which no words could be distinguished. Then these sounds died down and Galois remained motionless, rigidly embracing the coffin in which his father lay.

His mother knelt and tried to raise him. He was carried home. His head burned and he was put to bed.

The doctor came and told Evariste's mother, "He is a very sensitive boy. It is a misfortune to be as sensitive as he is. He will be better in a day or two. But he ought to have a quiet, peaceful life."

3 : 1829

Evariste sat in M. Richard's study. His face was thin and pale, his eyes were without fire, and the angle formed by his chin seemed more acute now than it had a few months before. M. Richard smoked his pipe and Evariste looked with empty eyes into empty space. M. Richard broke the silence.

"I know how you feel. I'm sorry—the only consolation I can give is the platitude that time heals all wounds. Like many platitudes, it's true. And for you there's something else that may help: work. You are a mathematician and you will do mathematics whether you decide to or not. It's stronger than you are. Why not accept this fate and throw yourself willingly into work? It may bring you peace; it will quicken the flow of time which, to repeat the platitude, heals all wounds."

Evariste did not answer. He sat as though he had heard nothing.

M. Richard asked, "What happened to the paper you sent the Academy?"

Evariste answered apathetically.

"The paper I sent the Academy? Yes! I have news about that paper. It's very, very funny. Here it is: one free afternoon I didn't know what to do and I wandered through the streets. I found myself in front of the Institute. I went in and asked the clerk what happened to my paper. He could hardly trace it. Indeed I almost began to think I'd never sent it. But finally he found a note about it. M. Fourier, the secretary, had sent it to M. Cauchy and it had never been returned. 'Are you sure it was not returned?' I asked. The clerk answered, 'Oh, absolutely. M. Cauchy sends so very few papers which are not his own manuscripts that I would certainly notice it.' Then he suggested that I go to M. Cauchy and ask him whether he had received my manuscript and what he had done with it. The clerk was very charming. He smiled and thought the whole thing was a very good joke. I didn't see what was so funny about M. Cauchy not sending back the manuscript. So I went to M. Cauchy's house. A woman opened the door.

Perhaps it was his wife, perhaps his servant. I asked politely, 'May I see Professor Cauchy?' The answer was, 'M. Cauchy is very busy, he can't see anyone.' Then I said that I'd like to have the manuscript I sent to the Academy and the Academy sent to M. Cauchy. She went away, closing the door in my face, and I waited. Then she came back and asked my name. I told her and she went again to M. Cauchy, again closing the door on me. Then she appeared with the final verdict from which there was no appeal. She stood before me and recited sharply, 'M. Cauchy knows nothing about a manuscript by M. Galois, doesn't have it, and doesn't recall receiving it.' This, M. Richard, is the end of the story of a paper sent by a young mathematician, Evariste Galois, to the Academy in the hope that it would be read, commented upon, and make its author famous."

M. Richard puffed at his pipe in silence. Then he said, "This is very bad news indeed."

Then after another pause, "Tell me frankly, Evariste, whether you are sure of your results, whether you believe in their correctness and importance."

"All right, M. Richard! I'll answer you, perhaps more frankly than you expect. I'll answer you in a way I wouldn't have dared answer you two months ago. I believe that my results are correct and important. After writing them up and sending them to the Academy, I made further progress. I have new results. But there's much to be done. The field's immense. There are lots of things I don't understand. But someday maybe I'll find complete clarity. I think I'm on the track of the greatest discoveries in algebra of this century. I think a new algebra will start from my work. But there aren't many people in the world who can appreciate what I'm doing. M. Cauchy could, if he'd take the trouble. And M. Gauss, yes, he'd know how important my work is."

M. Richard looked puzzled. He went to the bookshelf, took out a volume and handed it to Evariste open at page sixty-five.

"Here's something that will interest you. Here is a mathematician who would be interested in what you are doing. This is Niels Henrik

Abel's paper. As you see, it appeared four years ago in a new German journal." He pronounced the title pedantically: *"Journal für die reine und angewandte Mathematik."*

Evariste grasped the volume, translating the German words slowly and clumsily.

Proof of the impossibility, in general, of solving algebraic equations of degree higher than the fourth. It is known that generally one can solve equations up to the fourth degree. But, if I am not mistaken, no satisfactory answer has been given to the following question: Is it generally possible to solve algebraic equations of a higher degree? This paper answers the question.

Evariste quickly turned one page after another. His eyes lit up, his cheeks became less pale. Unconscious of M. Richard and of the surroundings, he exclaimed, "I see! Of course. Very interesting! Very interesting, indeed!"

When he reached page eighty-four, he translated the conclusion:

It is impossible to solve algebraically an equation of the fifth degree. It follows from this theorem that it is impossible to solve a general equation of degree higher than the fifth. Therefore in the general case, only equations up to the fourth degree can be solved algebraically.

He closed the book. His apathy was gone and he asked excitedly, "Where is Abel? Who is he? By now he may be on the right track. Perhaps he has found the general conditions for solvability, too. I want to see him or write. He'll understand how important my problem is, and how difficult. Where is he now? How old is he? He must be young."

M. Richard said calmly, "I shall tell you in a moment all I know about Abel. But first I want to show you another of his papers that just came."

M. Richard handed Evariste a recent issue of Crelle's *Journal.* Galois read the title, *On a Particular Class of Equations Solvable Algebraically,* and quickly looked through the paper, written in French. His excitement grew.

"It's obvious. It's the same direction. The paper was written in March, 1828. Then he didn't have the results I have, but by now he may know the solution. He's a great mathematician. I must meet Abel. Please tell me where he is. I want to write to him immediately. Here it says Christiania. Is he there?"

M. Richard answered, "Abel is dead. Only by chance I know his tragic story. He died of consumption in Norway a few months ago. When he died in April, in utter poverty, a letter was under way with an offer of a university chair in Berlin. The letter never reached him."

"How old was he?"

"Twenty-seven years old. There's something more in his story that will interest you. He had sent the manuscript of an important paper to the Academy and it was sent to M. Cauchy. No one knows what happened to the manuscript."

Evariste's eyes dilated with anger and hatred. "Abel died in poverty at the age of twenty-seven. His manuscript was lost by M. Cauchy." The boy spoke insistently. "These aren't isolated incidents. They form a pattern. Don't you see, M. Richard, that they're connected? There's my father's death, there's the rebellion at Louis-le-Grand, there's the vanishing of Abel's manuscript and mine, there's Abel's death. They look like isolated incidents with no relation between them. They're entirely different, on different planes, in different places, they concern different people. They range from Norway to Paris and to Bourg-la-Reine. But, believe me, M. Richard, they are not isolated. They are connected with each other and with millions of other events. They form a pattern, a clear pattern.

"And the connecting link is the malicious social organization under which we live. It killed Abel because it has contempt for the poor and hostility toward genius."

Evariste raised his voice and M. Richard looked uncomfortably at the walls of his room as though to test whether they were thick enough to screen the sounds made by his visitor.

"A malicious social organization makes genius go unrecognized in favor of fawning mediocrity. I know that very well. But I know more. I know the brutal, ruthless force of this malicious social organization."

Evariste stopped. M. Richard felt relieved when he heard Evariste's voice become calmer and more matter-of-fact. But with each sentence it began to rise again until it became a loud and uncontrolled torrent of words.

"The same force that killed Abel poisoned the mind of M. Cauchy so that he has no kindness toward others, no interest in other human beings. It's the same force against which the students rebelled and which threw more than a hundred of them out of this school. This force killed my beloved father. The parish priest was only an instrument. An outside force sent him to Bourg-la-Reine with distinct orders to undermine and destroy my father's authority. This force is responsible, and not the priest who forms only one small wheel in the machinery of tyranny and suppression. It is this force that I shall have to fight. I tried to escape from this fight by doing mathematics. But this force has invaded my life and taught me you can't escape. Individuals aren't responsible; the rotten social system makes them act that way. This is what my father taught me. I didn't see it before my father died, but I see it now."

M. Richard's amazement grew. When others told him that Galois was strange, M. Richard thought they would say this about anyone who had great mathematical talent. But now he saw that his guest was in fact strange, and that this strangeness seemed to have nothing to do with mathematics.

"Do you know, Galois, that you are talking like a Republican?"

"I know."

"You can't mean what you say. I know our world isn't the best possible. Progress is slow and painful, often followed along a path that seems to go backward. But we do achieve progress! Now we have peace. We have a charter that grants reasonable rights to the people. Everyone who really wants to work can. Any disturbances, any revolutions, would only bring back terror, increase misery and poverty. Of course tragic things happen, but they're often only accidents. If he hadn't been sick, Abel would now be a professor in Berlin. And consumption attacks rich and poor alike. If the new parish priest hadn't come to your town, your father would still be alive. There are good and bad priests just as there are good and bad mathematicians. M. Cauchy is a strange man who writes a paper every five minutes and doesn't have time for anything else.

"Obviously, they're all accidents. We ought to think not about destruction, but about construction. If I teach well, if you do mathematics well, then we are two wheels in their proper places, working correctly. If all the wheels work well, the whole machinery will function properly. But if I stop teaching, if you stop doing mathematics, we upset the machinery. But what you propose is much worse. You'd like to join those who want to smash the whole machinery to pieces. You'd achieve chaos and terror; you'd free the forces of cruelty and brutality. Compared with them our present world would seem a peaceful dream full of idyllic beauty."

Both M. Richard and Galois felt a wall growing between them. It grew thicker when Evariste replied in rising anger:

"You talk about orderly machinery and about smashing it to pieces. It looks like a good comparison but it isn't. There is no machinery! There's only a pile of rusting junk. The best, the vastest material, the people who are born poor, what part do they play in your machinery? They rot and rust in idleness and wretched misery if they can't find work. They rot through overwork and exploitation if they're lucky enough to find someone who will graciously accept their sweat and toil for a morsel of bread. Where do you see, M. Richard, any sense, any rhythm or design in this machinery? By God, a few years more and it will not need much of a revolution. It will disintegrate by itself, decompose and stink to heaven. The sooner we begin to destroy it the better for the future of the world."

They both felt that these words would always stand between them. Galois thought, "I came here for consolation and I'll carry away one more disappointment. How can a man who is a good teacher, whom I thought understanding and intelligent, understand so little? How can he believe that this is a world worth living in? How can he fail to see its horror and injustice? And I thought he was my friend!"

M. Richard thought, "He's young, but even his age and tragic experience aren't enough excuse for what he says. He ought to have more sense than to pour out this dangerous talk in my study, here at Louis-le-Grand. He ought to keep his thoughts to himself. They are subversive and dangerous."

M. Richard was anxious to end the conversation. He wanted to do it as graciously as he could.

"I think that we'll never agree on this point, and I think there's no sense in prolonging our discussion. After all, there's no reason why we must have the same views. We had better avoid this subject altogether. But there are many other problems we can discuss. Above all, there is one thing I would like you to know. I firmly believe that your great task is to do mathematics. It would be unfortunate if you were to neglect this task."

When Galois went back to his dormitory, he felt not merely depressed—he felt angry and he despised himself. Again and again he went over the details of his conversation with M. Richard and, biting his lips, repeated to himself, "Why did I say all that to him? I was a fool, a complete fool."

The teachers at Louis-le-Grand had their last opportunity to write remarks about Galois. Here is how they used it:

> His conduct is very good at intervals, but sometimes it is very bad. His facility in learning science is known. When working, he is occupied with it exclusively and rarely does he waste his time. His progress is proportional to the extent of his abilities and to his interest in the sciences. His character is strange and he pretends to be more strange than he really is. His behavior during religious services is not always as good as one would desire. His health is good.

For the last time M. Richard wrote his remarks about Galois. After the first term he had written, "This student has a marked superiority

over all his colleagues." These were words of praise that M. Richard had never used before. Again after the second term he had written, "This student works only on the most advanced parts of mathematics." But when the third term came, he wished to turn the page quickly and to forget Galois. He put down mechanically what he usually wrote for every good student, "Conduct good, work satisfactory."

In class competition, Galois won the first prize in mathematics, as everyone had expected. M. Richard had also hoped that Galois would win the first prize in the general competition. This would have meant —among other honors—entrance into the Polytechnical School without examination. He did not receive the first prize, but only the fifth. The examination problem was not very difficult and several students gave perfect solutions. Galois' was too brief, the reasoning too concisely presented. Another student won; his name was Bravais, and in due time he became a professor at the Polytechnical School and a member of the Academy.

4 : 1829

M. Dinet had been an examiner at the Polytechnical School for the past twenty years. During examination weeks he worked nine hours a day, until he had examined the few hundred students who hoped to make the grade. Some ten years ago he had had a nervous breakdown because of the sickening repetition of questions, and the doctor had prescribed that he leave Paris for a few months. Then he recovered, came back to repeat his old questions, to listen to his own annoying voice. The answers were still more annoying, for to these he *had* to listen. After two minutes, nay, after one minute, M. Dinet knew whether the candidate was worthy of admission to the Polytechnical School, from what book he had studied and how much he understood of it. But M. Dinet prided himself on being decent and prolonging the examination for the student's sake, though with infinite boredom he anticipated the exact formulation of the second and third sentences even before the first was finished. The only way to stop the flow of words was to interrupt the preconceived line of answers with new questions; but there was little fun for M. Dinet in replacing the student's voice by his own.

It was a hot day. M. Dinet was perspiring, tired, thirsty, and he wished the day were over. His right shoe pressed unbearably on a corn, and he longed for his armchair and slippers. But there were still three candidates to be examined. The janitor was just washing the blackboard after the previous student, when M. Dinet, drumming with his fingers on the desk, tried to suppress a yawn and said, "Next candidate, please." Then without raising his head: "Name?"

"Evariste Galois."

"Tell me what you know about the theory of logarithms."

M. Dinet closed his eyes. He knew what would come. He would hear that $b = \log_a c$ if $a^b = c$. Euler had used these letters in his book on algebra, and since then every student had used them when talking about logarithms. Then he would hear that the logarithm of a product equals the sum of the logarithms.

"How terrible! How awful! Oh, infinite boredom! Twenty minutes more and I shall finish the examination of this—what's his name—and then two others. Then my slippers—oh well, let's listen."

But there was nothing to listen to. Something was wrong. M. Dinet was glad, it might mean a new experience. Perhaps a deaf and dumb student was trying to pass the entrance examination! That would be interesting. At least the student could write. He heard the chalk colliding with the blackboard. He would have to look. He raised his head, heavy with sleep, and saw on the blackboard:

$$1, a, a^2, a^3, \ldots .$$
$$0, 1, 2, 3, \ldots .$$

M. Dinet felt less sleepy. He became interested. This was something new!

"Would you have the goodness to explain what you are doing?"

A flat voice recited apathetically, "These are two progressions; a geometric and an arithmetic progression. The terms of the arithmetic progression are the logarithms of the corresponding terms in the geometric progression, and *a* is the base."

"Very good," said M. Dinet. He waited for the voice to continue. But the encouraging "very good" did not quicken the flow of the student's words. He merely added, "And so on," thus wiping out a great part of the good impression.

M. Dinet asked impatiently, "What do you mean by 'and so on'? What is the next step?"

He waited for a while.

"Young man, I cannot draw answers from you by force. Either you wish to answer or you do not."

Galois experienced the same emotions that he had felt so many times before: the rising anger, the burning of his skin, the strain of the effort by which he tried to suppress his anger. His face became red, his voice choked, but his answer appeared calm and apathetic.

"Between every two numbers of the geometric progression one may insert $(n-1)$ numbers and the same between two numbers of the arithmetic progression. Then the numbers of the arithmetic progression are still logarithms of the corresponding numbers in the geometric progression."

"You must make yourself clearer. What kind of numbers do we insert?"

Galois looked with disdain at M. Dinet. The thought that there was anyone who could judge whether he, Galois, was prepared for the Polytechnical School was hard to bear. But the thought that it was M. Dinet was doubly unbearable.

"It is all quite clear. If one inserts $(n-1)$ numbers so that the respective progressions remain geometric or arithmetic, as I clearly assumed, then everything is uniquely determined and there is nothing further to add."

"It may be clear to you, but it may not be clear to me. Pray write down these expressions; otherwise we might just as well terminate our conversation."

Galois, without saying a word, wrote on the blackboard:

$$1, a^{\frac{1}{n}}, a^{\frac{2}{n}}, \ldots\ldots a^{\frac{n-1}{n}}, a;$$

$$0, \frac{1}{n}, \frac{2}{n}, \ldots\ldots \frac{n-1}{n}, 1.$$

M. Dinet looked up and sighed with relief. He thought, "What manners, what manners these young men have nowadays! I don't like him. I shall wipe off his supercilious expression, if it's the last thing I do today." Then he asked, "Can I now insert $(n-1)$ numbers in one interval and $(m-1)$ numbers in another, where n is different from m?"

"By all means, sir " said Galois.

"Therefore, can the number of terms vary from interval to interval?"

"I said by all means, sir."

"Can you explain why?"

Galois knew by now that only irony would conquer the rising wave of his rage.

"Is it not obvious to you, sir?"

M. Dinet gesticulated excitedly.

"Suppose, sir, that it is not. Suppose that I wish you to explain it to me. And suppose also that I tell you that if you do not succeed in explaining to me this small and trivial matter, you will fail your examination. What then, M. Candidate, would be your answer to my question?"

Evariste looked into M. Dinet's eyes. In his right hand he mechanically squeezed a sponge. By now neither persuasion nor irony could control his mounting anger. It became stronger than he was. It even distorted his vision. It changed M. Dinet's face curiously. He became much thinner, his features sharper. M. Dinet now looked like the parish priest of Bourg-la-Reine. Yes, he was the parish priest, only aged; his features became still sharper and more grasping. It was the parish priest at whom stones had been thrown by those who loved the mayor. Yes, it was the parish priest. An oppressive fog spread over the room. If it would only lift, he would see men throwing stones at the parish priest who now sat at the table indifferent to the wrath of the people around him.

A shrill voice dispersed the fog.

"I repeat: what would be your answer to my question?"

Galois raised the sponge and threw it at M. Dinet's head. It hit precisely where it was intended to hit.

He cried out joyously as though he were relieved of the greatest burden of his life, "This would be my answer to your question, sir."

Then, without looking back, he went out and closed the door behind him. He knew that he had closed it forever.

V

IN THE YEAR OF THE REVOLUTION

1 : 1830

IN FEBRUARY, 1830, Galois officially entered the Preparatory School. This was a weak and humble copy of the Normal School which had been founded during Napoleonic times and suspended during the Restoration. In 1826, four years after the Normal School was closed, the Preparatory School was opened to turn out masters and professors for the royal colleges. The school was housed in du-Plessis, formerly a part of Louis-le-Grand. Between the Preparatory School and Louis-le-Grand there was not only proximity in space, but also proximity in spirit—the same discipline and the same supervision. Only the teaching level was higher and the specialization greater.

To be accepted for the Preparatory School, Galois had first to become a Bachelor of Science and then to pass the entrance examination. He succeeded in doing both.

His examiner in mathematics, M. Leroy, gave him eight points out of ten and wrote:

> This student is sometimes obscure in expressing his ideas but he has intelligence and shows a remarkable spirit of research. He communicated to me some new results in applied analysis.

M. Péclet, the physics teacher, wrote about Galois:

> He is the only student who answered me badly; he knows absolutely nothing. I have been told that this student has mathematical ability; this certainly astonishes me. Judging by his examination, he seems of little

intelligence, or has hidden his intelligence so well that I found it impossible to detect it. If this student is what he seems to be, I doubt whether he will ever make a good teacher.

Poor M. Péclet! How often has this note of yours, never designed for publication, been quoted as a glaring example of human stupidity and a monument to the blindness and folly of a teacher!

In the same year, 1830, three of Galois' papers appeared in the *Bulletin de Férussac*. In April there was a short note: *Analyse d'un mémoire sur la résolution algébrique des équations*. Then in June, an equally short note (two pages): *Sur la résolution des équations numériques,* and a longer paper (eight pages): *Sur la théorie des nombres* accompanied by the following note, "This paper forms a part of the researches by M. Galois on the theory of permutations and algebraic equations."

These notes contained only fragments of Galois' results, some of them merely stated without proof. The theory was more fully formulated in a paper which he sent in February to the annual competition for the yearly Academy award. This time he had no illusions; he did not dream of triumph and success; but he knew that if recognition did not come, the disappointment might be his, but the final humiliation would be the Academicians'.

It was not easy for Galois to return once more to the atmosphere of Louis-le-Grand that prevailed at the Preparatory School. But he was also attracted by this hated atmosphere; the bonds of hatred can be as strong as those of love and devotion. The school, the Academy, were the battlefields where he had been humiliated and where he now had to return and fight again. But before the year was over Evariste saw a broader battlefield and a more important fight. The battlefield was Paris, the fight was for the rights of the people of Paris, of France, and of the whole world.

Since Prince de Polignac had become the King's first minister, the bourgeoisie of France had lived in dread and expectation of a revolution. They abhorred the noblemen who humbled them by their superior manners and impeccable taste. They hated the clergy because it stood where the noblemen stood. They hated the King who represented both the nobility and the clergy.

The King did not believe in concessions. They had not saved his brother. Louis XVI had made concessions and retreated. He had retreated again and again until further retreat was cut off by the knife of the guillotine. Charles X thought the people of France could only be ruled by a strong hand. To them concession meant weakness and retreat meant cowardice.

On the second of March, 1830, the chambers, assembled in the *Salle des Gardes* at the Louvre, were to listen to the speech from the throne. Early in the morning all places reserved for the public were occupied, and eager crowds waited outside the palace. At one o'clock Charles arrived. Everyone stood up as the graceful King, dressed in a general's uniform, moved toward the throne. This perfect actor lost his balance for a moment while ascending the steps, clumsily covered with rich rugs; the two-cornered hat fell from the King's head, rolled down to the feet of the Duc d'Orléans, who picked it up eagerly and handed it to the King above. This trivial accident and its deeper symbolic meaning was soon discussed and analyzed over the whole of France.

The King's speech was long and boring. Everyone waited tensely for the expected fireworks. They came at the end.

"Peers and deputies! I do not doubt that you will support my endeavors to carry out this good work. Should culpable machination raise obstacles in the path of my government . . ."

Charles looked up from the script, sharply eyed the audience on the left and, stressing each word, continued, "which I refuse to anticipate," then he looked at the unrolled sheet again and read, "I shall find the strength to overcome them in my determination to preserve the public peace, in the just confidence of the French, and from the love they have always evinced for their king."

This was an open declaration of war upon the chamber where the

Liberals held a majority. A few days later, the chamber, where the ultras were in a minority, struck hard and swiftly, so as "not to allow the folly and ineptitude of a few men to destroy liberty."

The majority of the chamber in the famous address of the 221 replied, "The charter requires harmony between your wishes and those of your people. Sire! Our loyalty and our devotion force us to tell you that such harmony does not exist."

The King in his palace listened to these words, while playing with a piece of paper and pretending boredom. Then he said that his decisions were unalterable and dismissed the frozen deputies.

The ultras boasted, "These people did not know what a king is; they know it now; a breath dispersed them like chaff."

But the 221 deputies boasted, "Never did a sovereign's crown, not even that of Louis XVI, receive such a challenge."

The King dissolved the chamber. All the political overtures were now played out. At any moment the curtain might rise on the last act of the Bourbon play. Its possible actors, contents, finale, were discussed endlessly in the streets, wineshops, cafés, in the Napoleonic clubs, and in the small but active and always conspiring Republican clubs. The students, and especially those of the Polytechnical School, responded to the rhythm of political events.

Laplace, who was supposed to have reorganized the Polytechnical School so as to suit the Bourbons, had been dead for three years, and did not live to see the futility of his efforts. The Polytechnical students conspired while playing billiards, while preparing their lessons, while fencing or eating. But there was no spirit of rebellion in the Preparatory School. There among the fifty students only one was peculiar; instead of preparing for examinations he disturbed and bothered his classmates with idle talks about Charles, Bourbons, Jesuits, liberty, and tyranny. He was not only strange and affected, but seemed to take pride in being different from others. When he answered questions in mathematics, he did it as though he were asleep or bored to death. Or he assumed a ridiculously suffering expression while obviously rejoicing (so his classmates claimed) at being the only one who knew the right answers. He annoyed his fellow students by writing meaningless formulas on small pieces of paper and pretending to be deep in thought, deaf, and superior to the world around him.

During one of these hectic days, just after Charles dismissed the chamber, M. Leroy came to the mathematical seminar looking particularly grave. He announced to his twenty students that he had something interesting to say. Evariste thought that perhaps M. Leroy was human, after all, and that perhaps he wished now to make a confession of his political faith. But what M. Leroy said was that Sturm had stated an interesting theorem in algebra. He quoted the theorem, but regretted that the students would have to wait for Sturm's paper to learn the proof. Then, glancing at his small audience, he saw an ironical smirk on a face he knew well. The professor focused his eyes on that face, and the sarcasm in his voice was slight, hardly traceable.

"Good that you are here today, M. Galois. Perhaps you will be able to help us."

Evariste did not answer, but his ironical smile vanished. Everyone looked at his now tense face, and students whispered to each other.

"He will find it."

"No, he won't."

"His brain will crack."

Suddenly Galois' eyes lit up. He went to the blackboard and wrote out the proof. Some of the students diligently took notes, copying the symbols with which Evariste covered the blackboard. Among those who now watched Evariste, there were only two who did it without hostility or jealousy. Both were students of the second year. One, Bénard, Evariste's cousin, took some family pride in Galois' performance. The other was Auguste Chevalier, with the pink fat face of a cherub, clumsy, shy, and as lonely as Evariste among his classmates. It was Auguste's misfortune to have deep religious beliefs, and the less they were shared by others, the more convinced he became that it was his duty to do missionary work among the savages of the Preparatory School.

While dreamily copying Evariste's mathematical signs from the blackboard, Auguste was suddenly struck by an inspiration.

"He is a genius! For the first time in my life I see before me a real genius! Eighteen others look upon him with misgiving and jealousy. I am the only one who knows that it is a great privilege to witness the work of this genius. I see because I was taught what love means and what genius means, because my eyes were opened by my beliefs."

Having copied the symbols from the board, Auguste wrote in a neat, careful handwriting, "Evariste Galois is a genius. Must become Evariste's friend. Will try to convert him to Saint-Simonism."

2 : Sunday, July 25, 1830

On Sunday, July 25, the ministers assembled at St. Cloud to sign the ordinances that would suspend the constitution of France, dissolve the chamber, and wipe out the freedom of the press. In silence they took their places around the table. Charles X had Prince de Polignac on his left and the Dauphin on his right.

Baron d'Haussez asked de Polignac, "How many men have you in Paris?"

"Enough to suppress any rebellion."

"Have you at least thirty thousand?"

"More than that. I have forty-two thousand."

Prince de Polignac threw a paper across the table to Baron d'Haussez.

"Why, what is this?" asked the Baron. "I find here only thirteen thousand men accounted for. Thirteen thousand men on paper means only seven or eight thousand fighting soldiers. Where are the other thirty thousand men?"

"The rest are quartered near Paris. If necessary, they can be assembled in ten hours in the capital."

De Polignac asked one minister after another to sign the ordinances. When the document came to d'Haussez, he took the pen but hesitated.

"Do you refuse?" asked Charles X.

"Sire! May I be allowed to address one question to Your Majesty? Is Your Majesty resolved on proceeding if your ministers draw back?"

"Yes," said Charles X firmly.

Baron d'Haussez, Minister for the Navy, then signed. Prince de Polignac looked around with triumphant eyes.

The King said, "I count upon you, gentlemen, and you may count upon me. Our cause is one. It is a matter of life and death to us."

He rose and walked up and down the room, his every gesture that of a king. And now he felt that he was a king.

On Sunday, July 25, Auguste Chevalier and Evariste Galois sat in the Luxembourg Garden. They had often spent their free afternoons together since they met at the mathematics seminar. But this was the first time that Chevalier tried to confide in his younger friend.

"You see, I stand with my older brother in political and social matters. He has always had a great influence on me. He is one of the disciples of the Comte de Saint-Simon. Have you heard about Saint-Simon?"

"Not much. Tell me."

"It is from Saint-Simon and from my brother that I first learned to admire science and especially mathematics."

"Why? What has Saint-Simonism to do with mathematics?"

"Saint-Simon's first book would answer your question. Read *Letters from an Inhabitant of Geneva to his Contemporaries.* In this book, Saint-Simon plans a general subscription, to be opened before Newton's tomb. Everyone would contribute—everyone, rich or poor, man or woman, each according to his means and his inclination."

"And what would happen then?"

"Every contributor would write down twenty-one names: three mathematicians, three physicists, three chemists, three physiologists, three writers, three painters, and three musicians—altogether twenty-one."

"The mathematicians come first."

"Yes, they are on the top of the list. Then the twenty-one who received the most votes would be called 'The Council of Newton.' All the money collected by subscription would be given to this council, and one of the mathematicians would become its president."

"Again the mathematician would be the first among the first."

"Yes! You see the appreciation of mathematics in the early ideas of Saint-Simon. The council, under the leadership of a mathematician, will be the spiritual government of the world and unite all the nations into one great nation."

Galois was amazed that his friend took this fantasy so seriously. He asked cautiously, "But do you think such a plan is sensible, and can be realized?"

"I know it sounds fantastic and perhaps even ridiculous if analyzed logically. Try to analyze *Hernani* logically—it will seem a collection of nonsense, but it's the greatest drama of our century. Saint-Simon's early work may seem unrealistic, but it's important, and it led to the present much more practical plan of the Saint-Simonists."

"And what does Saint-Simon believe in now?"

Chevalier was astonished by Galois' ignorance. He explained patiently, "Saint-Simon, from one of the noblest French families, died in poverty five years ago. To his disciples—one was my brother Michel—he said before he died, 'The fruit is ripe; be it yours to pluck.' "

Galois did not feel embarrassed that his lack of knowledge had been exposed. He asked indifferently, "Then what do his disciples believe now?"

Chevalier answered with the calm and sweetness of an evangelist, "We believe that love will conquer the world and that hatred will cease to exist. There will be no more competition, no hereditary possession, and no war. Brotherly love for all mankind will triumph and a new Christianity will be established."

"How will you do it?"

"By spreading our beliefs, by preaching love, by giving power to the best, to the most capable, by rewarding all according to their works. Our slogan is: 'To each according to his capacity; to each capacity according to his works.' "

Galois repeated the sentence.

"To each according to his capacity; to each capacity according to his works." Then he spoke excitedly. "Don't you see there's a great contradiction in your philosophy? You'd like to conquer the world by love. But at the same time you'd give power according to capacity. Suppose we accept this plan. Then you'd regard men according to their works. Where is your love for the feeble, for the idiot and the sick, for the most unfortunate of this world? Don't they need food, shelter, care, warmth, even if their capacity is small? What about their needs?"

"For them there will be the charity that springs from love."

Galois interrupted violently, "Charity! How I hate that word! Charity, that makes the poor and unhappy man dependent on the good impulses of the rich and kills the will of the poor to fight against the rich. Charity, that replaces the sacred duty of the state by the whim of individuals. There are thousands of families here in Paris who eat bread that can be split only by an axe, and eaten only after soaking in water for two days. Their rooms have earthen floors covered by straw; the air is foul and damp, and it's dark on the brightest day. To them you'll bring charity and love. By God! They ought to hate. It's their right to hate, to destroy people who think their condition natural.

"Yes, love sounds beautiful. But love can rule only after an eruption of hate that will shake the world to its very foundations. Love can grow only on the ruins of the old world. Only hatred can destroy that world. The revolution didn't succeed in doing it; someday the people will have to try again."

Chevalier was frightened by the passion and fire with which Galois spoke; he did not dare to prolong the discussion and only added evangelically, "I thought you cared only about mathematics."

"No! Mathematics isn't my only concern; but, I'm sorry to say, so far I haven't done anything else. I've been living in emptiness as though I were afraid to touch the earth. But some day you will see that mathematics isn't the only thing I care about."

He remained silent, hesitating whether to say what he wanted to.

say. Then, as though he were confiding his greatest secret to his worthy friend, he repeated his father's last words:

"Even mathematics, the most noble and abstract of all the sciences, has its crown in the air but its roots deep in the earth on which we live. Even mathematics will not allow you to escape from your sufferings and those of your fellow men."

Then he whispered:

"If I knew that a corpse would stir the people to revolt I would give them mine."

3 : Monday, July 26, 1830

The suburbs of Paris were quiet. The ordinances were no concern of those who worked for fourteen hours daily and smelled of sweat and dirt on this hot July day. They did not read the newspapers and cared little for the freedom of the press which was now threatened. In the Chamber of Deputies they had no representation and were little concerned that its existence was now in danger.

Writers, editors, and managers of the many Parisian journals assembled in the offices of the newspaper *National* on rue Neuve-Saint-Marc. M. Thiers, the brilliant young editor of the brilliant journal *National,* read aloud from a paper to an audience of about fifty. Since he had a perfect sense of history, he was quite conscious of making it right there; since he was a fine actor, he knew how bad it is to overplay one's part. Thus he seemed calm, composed and dignified as he read the document in his Marseilles accent and in the dry tone of a lawyer who presents a case in which he is not personally involved.

"In our present situation, obedience ceases to be a duty. The citizens, who before all others are called upon to obey, and the writers for the public journals ought to be the first to set the example of resistance to that authority which has divested itself of the character of law."

In a slow, monotonous, but distinct tone he read the draft of the journalists' manifesto to its closing sentences:

"The government has this day lost the character of legality which commands obedience. For ourselves, we shall resist; it is for France to judge how far the resistance should extend."

Then, putting aside the paper, he added less dryly, "The one thing we can and must do is to protest against the attack on the freedom of the press. I move that we sign this manifesto and make the protest known to France. You know that by doing this we risk much. But to do nothing means to risk even more; it means to risk the good name of the press."

There was a long and stormy discussion. Late at night, forty-five men signed the journalists' manifesto. This document played the part of a pebble which on its journey down the snowy mountain forms the nucleus of an avalanche.

A postilion traveling to Fontainebleau on the night of July 26 told one of his comrades the news of the ordinances.

"The Parisians were in a fine stew yesterday evening. No more Chamber of Deputies, no more journals, no more freedom of the press."

The other replied, "What do I care as long as we have bread at two sous and wine at four?"

When this anecdote was repeated to Prince de Polignac, he remarked philosophically, "The people care about only three things: work, cheap bread, and low taxes."

In this the prince erred.

4 : The Glorious Tuesday, July 27, 1830

On July 27, most of the Parisian newspapers appeared, but not all. Some copies of the *Globe,* where the journalists' manifesto was published, found their way into the Preparatory School and into Galois' hands. Exultantly he read the words defying the King's order. When he saw the signatures of these forty-five brave men, Galois thought, "This is the first spark. It came suddenly—sooner than I expected. But will it change into a raging fire? Fire! Powder! Barricades!"

The thought possessed him. He heard the whistle of bullets, he smelled powder, he felt the recoil of a musket. He created scenes in which he spoke to the people, to the people of France, always moved by noble words and noble deeds, always ready to die for the cause of liberty. He led them, he fought with them, and he saw victory near. Then suddenly a bullet would hit him. He would die on the barricades of Paris.

"Galois died like a hero."

Yes, he heard someone calling his name. He felt the hand of a giant who lifted him up and then let him fall from the heights of the tower of Notre Dame to the level of the Preparatory School. He bounced a few times while falling, but landed safely just in time to hear the words of the tutor.

"Galois! How are you getting along with your studies?"

"Sir, did you read the journalists' manifesto?"

"Don't you think, Galois, that it would be much better for your own good to busy yourself with the approaching examinations than with the journalists' manifesto?"

"No, sir! Just the contrary. I believe that it would be much better for me to busy myself with the journalists' manifesto than with the examinations."

"In that case, Galois," said the master with an air of finality, "you had better discuss the matter with M. Guigniault. You may even be able to convince him."

"I shall be very happy to try, sir."

Galois felt strong. Before, his hatred of the school had been mixed with fear. But now the fear was gone and only hatred remained. The masters, the professors, and even M. Guigniault, seemed to him small and insignificant. He, Galois, had behind him the strength of the people.

When Galois was called before the director, he looked calmly into the big bony face and joyfully anticipated the show of his own strength.

But M. Guigniault did not seem eager to start the conversation. He looked through Evariste into space while his fingers played with a heavy silver chain hanging on his black waistcoat.

"M. Haiber has reported the conversation you had with him. Do you admit that what I heard is true?"

"Absolutely, sir."

"So you admit it. Do you know that such behavior may lead to expulsion from our school?"

"There's nothing I'd like better today than to find myself outside this school on the streets of Paris, myself and all the other students."

"Thank you for being so frank. But we shall not allow you to go out; neither you nor the others. It is our duty to protect our students even if they refuse to appreciate or understand our actions."

Galois answered angrily, "Sir! I've heard this argument about the good of our students at Louis-le-Grand since I was thirteen. It's an old and worn-out story. Everything the school has done, does, and will do, is always for the good of the students; it does it unceasingly, night and day, until it makes them miserable and breaks their spirit. I wish the school would leave the care of my own good to me."

There were both disgust and hatred in M. Guigniault's eyes. Soon, however, he regained control of his anger.

"Galois! This conversation will lead nowhere. I'll be frank with you. I have many enemies who think I'm too liberal. I do not think the clergy were delighted when I was appointed director of studies in the Preparatory School. And I do not approve of the ordinances." He hesitated for a while. "I personally disapprove of the ordinances; I very definitely disapprove of them. But this is only my personal belief. As director of the school, I must steer it clear of politics. My purpose is to regain for the school the dignity and importance which it had when it was the Normal School of France." He hesitated again. "Perhaps you'll admit that you are not, and never have been, an easy student. We have kept you here because we believed in your mathematical ability. We believed M. Richard and M. Leroy more than the examiners at the Polytechnical School. I offer you a compromise. If you'll help me, if you will not try to stir up trouble here in the difficult days that may lie ahead, then I can promise you, on my part, to forget all about your conversation with M. Haiber and with me just now. Do you accept my proposal?"

"Thank you, sir, for being so outspoken. But I cannot accept your offer. I know I'm risking my future here, but speaking equally frankly, I don't think that's important. I'm sure the students of the Polytechnical School, the students of other schools, will be on the streets of Paris today or tomorrow. It's my task to bring the students of the Preparatory School into the streets of Paris."

The director became fatherly.

"Imagine you succeed. Let's imagine, absurd though it may sound, that revolution breaks out and some of your comrades are killed. How would you feel then? Wouldn't the thought that you had caused their death haunt you for the rest of your life?"

"No, sir! It wouldn't be I who caused their death, it would be the regime. They wouldn't have died for me, but for France and the liberty of the people."

In the cafés and restaurants of Paris, hundreds of copies of the *Globe, National,* and *Temps* were bought, read, and discussed. Everywhere on that Tuesday, similar brief and forceful scenes were enacted as though rehearsed and staged by an invisible producer.

Someone would dramatically read aloud the journalists' manifesto to his accidental audience, and then shout, *"Vive la charte!"*

Urchins, roaming the town in search of adventure, liked the sound of these words which they did not understand. But they put into them a meaning of their own. The cry promised an exciting spectacle. Thus they shouted more violently and persistently than the bourgeoisie:

"Vive la charte!"

Through them the cry spread into the suburbs. The unemployed and the families of workingmen also liked the sound of these words whose meaning they did not understand. But they put into them their own meaning. The cry meant a twelve- and not a fourteen-hour working day, a bed to sleep in, and bread for their children. Thus they shouted more violently and persistently than the bourgeoisie:

"Vive la charte!"

Soon the cry sounded all over Paris.

Until seven o'clock in the evening there were no riots and no fights.

At the Quai de l'École, near the Louvre, crowds were drawn up along the river walls. A man walked slowly through the multitude carrying a tricolor flag. No one moved, no one said a word, but some eyes filled with tears. Some men uncovered their heads and others saluted. They all looked with wide-open eyes as long as they caught a glimpse of the three colors in the light of the setting sun. They had not seen this flag for fifteen years. The silent man who carried the colored pennant had returned to them the vision of France's glory.

At the Place de la Bourse there was a wooden shanty used by a dozen soldiers for a guardhouse. Late in the evening, urchins, men, and women collected before the shanty, tossing the day's slogan at the soldiers:

"Vive la charte!"

The soldiers did not know what these words meant, but they disliked and feared them. They had been told not to heed any shouts and they obeyed. The urchins, outraged by the soldiers' calm, tore stones from the pavements and threw them at the soldiers. Still there was no response. Then one of the stones struck the chest of a soldier. In a rage, he fired without aiming. A woman fell. Kneeling beside her, a man took her hand and cried dramatically, "She is dead! She is dead! Murderers! Murderers!"

The man was tall and strong. He took the woman's body in his arms and moved, followed by the crowd, toward the brilliantly lighted Théâtre des Nouveautés. The procession forced its way into the theater and entered the orchestra just as an actor was gracefully bowing to kiss the heroine's hand.

A voice from the audience rang out, more dramatic than any on the stage:

"Stop the play."

The audience, the bowing actor, and the heroine whose hand was being kissed all turned toward the voice.

"Stop the play. Here in my arms I carry a woman's body. She was killed by a soldier's bullet because the people cried, *'Vive la charte!'* "

The actor moved to the footlights and, raising his clenched fist, repeated the words as though they were part of the play.

"Vive la charte!"

The audience chanted wildly, *"Vive la charte!"*

On the night of July 27, a manufacturer of the Faubourg Saint-Marceau said to his friend, an editor of the *National,* "Take care what you do. If you give the workmen arms, they will fight; if you do not give them arms, they will rob."

In this the manufacturer of the Faubourg Saint-Marceau erred.

5 : The Glorious Wednesday, July 28, 1830

The student Bénard said, "Galois is right! Our place isn't here, but on the street. Look out the windows, and you'll see the barricade on the rue Saint-Jacques and the students of the Polytechnical School. I don't know whether they forced their way outside or were let out by the school authorities. But you see that they are on the barricade. It's time to decide what to do. The door to Louis-le-Grand has been closed since yesterday. The door leading to rue du Cimetière Saint-Benoît is closed and watched. We're like prisoners here and we have no chance to escape. But we can force our way out if we do it together. I agree with Galois that our time has come and that we must storm our way out. You fear consequences, but. . . ."

Voices interrupted, "Nonsense! We aren't afraid. We don't want a revolution."

A voice was heard, no one knew whether in fun or in earnest, "Long live the Revolution!"

The speaker tried to continue. But the chorus of pupils was noisy and persistent. Finally the speaker made himself heard.

"Let's not be cowards. . . ."

"He calls us cowards."

"Take it back."

Again the speaker managed to become audible.

"You can convince me that you're not cowards by deciding to go out and fight."

"You're a coward yourself."

"Shame! Shame!"

Two students started simultaneously toward the rostrum: Galois and Bach.

"We don't want Galois."

The majority repeated rhythmically, "We want Bach, we want Bach. . . ."

The weak countercry of "We want Galois" was drowned out. Bach reached the rostrum first. He was the best student in his class and looked just as such a student ought to look: neat and pedantic.

With a smile, friendly and at the same time superior, he waited until the noise died down. Then he said smoothly, "Fellow students! We have discussed our problem long enough. I say 'we,' but in reality we have had to listen most of the time to Galois." (Applause and laughter.) "We are debating instead of working. As far as I can see, only a small group would like us to break out of school. There are some who would be ready to go out if it were not against M. Guigniault's wish. We can have enough confidence in M. Guigniault that he will act wisely and decently and that he will have our welfare at heart. Therefore I move that we invite our director M. Guigniault to our meeting."

"Give me the floor, please."

Galois went to the rostrum. He stuttered, his voice choked with emotion.

"Please listen to me. Don't laugh now! Don't joke now when the people's blood is flowing on the barricades. . . ."

"People's blood! What about the soldiers' blood?"

"It's the people's blood that is flowing in defense of liberty. It's the people's fight we must join. The people didn't ask the King to allow them to revolt. But you want to ask M. Guigniault whether he'll let us revolt. Don't you know what he'll say? If we know and still ask him, we're hypocrites who look for. . . ."

"Shut up Galois. Shut up, shut up."

The noise absorbed Galois's words, and only his gestures were visible. Suddenly he gave up, left the rostrum, and sank into the nearest empty chair.

The students then sent Bach to invite the director. When M. Guigniault entered, the students rose with deference and listened attentively to their master's oratory.

"Students of the Preparatory School! I wish first to express my most sincere appreciation of the confidence you have shown me by your invitation." (An oratorical pause) "We are experiencing grave days. I'm not afraid to say that I condemn the ordinances which curb the freedoms guaranteed to France by Louis XVIII, and that our King was ill-advised to dissolve the Chamber of Deputies and to sign the ordinances. I am on the side of the law!"

He looked at his audience, paused again, and then continued in a low, soft voice.

"But if you ask me whether we ought to support the Revolution, if you wish me to answer this question by a single 'yes' or 'no,' then I must refuse to answer. Here in this school we have a task which is above and beyond the shifting ground of political events. We must study, we must learn our subjects so that we can pass on effectively the knowledge our heritage gives us to the younger generation. This is the duty to France that we undertook and this is the duty to France that we must fulfill. To go out into the streets of Paris means to abandon that most sacred obligation."

Now the tone became fatherly.

"I want to convince you. I don't want to use force. I could have appealed to the police to keep order and compel you to stay inside the walls of the school, but I did not choose to do so. If you give me your word that no one will leave this building, then I can promise you that the doors of our school will be opened. I shall trust your word." (A transition to oratory) "Let us remember in this grave hour that suffering and tragedy oppress both sides. It is true that the people, ready to fight, are defending the freedoms threatened by the ordinances. But we must remember that soldiers are human too. They have taken an oath to the King which they wish to honor."

The speech was reaching its climax.

"If we try to achieve this higher level of understanding, then we must look with pain and sadness at the struggle that lies ahead. In this grave hour our task is clear: we must resolve to do our utmost to heal the wounds of France after the struggle is over."

Applause broke out. M. Guigniault waited patiently until it died down, then said, "Therefore, I ask you: Do you promise me not to try to leave the school until the struggle is over?"

A single "no" broke through the loud "yes" chorus.

"I regret that not all of you wish to give me your promise. May I ask which one of you refuses to give his promise?"

"Sir, I do."

Evariste caught a glimpse of Bénard's face turned toward the floor and of Chevalier's red cheek. The director looked at Galois with well-concealed triumph tempered by well-displayed patience.

"I'd like to compromise. I don't wish to use force or to invite the police. I therefore ask this one student who breaks the unity of our school: Can you at least promise me that you will not attempt to leave the school today or tomorrow?"

"No, sir," replied Galois.

"I shall go still further, just to show all of you how much I wish to use persuasion instead of force. Do you at least promise me that if you decide to go out, you will first tell me of your intention?"

"No, sir."

The director turned to the rest of the audience.

"You see clearly that I have tried to do my best. I am sorry that all of you will have to suffer for the unbelievable obstinacy of one student. But until this student changes his mind, the doors of the school will be closed and guarded. I regret this prisonlike arrangement as much as you do. But, after what you have witnessed here, I am sure not one of you will blame me. Before I leave you, I should like to thank you once more for inviting me." (Exit)

M. Carrel, a well-known writer, editor of the *National,* one of those who signed the journalists' manifesto, said to a Republican friend on the morning of July 28, "How can you believe in a revolution? Have you at least one battalion at your disposal?"

Looking around, he saw a man who was polishing his shoes with oil from a broken lamp. Pointing to this man he said, "That is a typical picture. That's what the people do—break the street lamps to polish their dirty shoes."

But the Revolution did come. It appeared on the streets of Paris, prepared by no one, organized by nobody, feared by the men who caused it, fought for by people who did not understand the slogan in whose defense they gave their lives. No one knows how or whence the first spark came. But on July 28, the fire of revolution was raging through the streets of Paris.

The tricolor floated from the tower of Notre Dame. The drums beat and the bells of Notre Dame rang, announcing to the world that the July revolution of 1830 was marching through the streets.

On Tuesday night the students of the Polytechnical School broke into their fencing rooms, grabbed the foils, broke off the buttons from their points, and sharpened the swords on the stones of the corridors.

When on Wednesday morning two hundred and fifty of them forced their way out of the school, they were greeted at the rue de la

Montagne-Sainte-Geneviève with shouts, *"Vive l'École Polytechnique!"*

One of the students, holding his cocked hat in the air, tore the white cockade from it and trampled upon it. Two hundred and fifty students angrily imitated this gesture amid wild cries of "Down with the Bourbons! Long live Liberty!"

A suspension bridge leads from the Island of Paris, on the Seine, to the City Hall. One hundred men marched towards this bridge to charge the City Hall, the nerve center of Paris. No cries, no slogans were heard, but only drums and the irregular beating of footsteps soon covered by the ever-increasing sound of rhythmically marching feet. A military detachment marched to the other side of the bridge, its fixed bayonets reflecting the bright July sun. Then suddenly, when the guard reached the bridge, its ranks opened and it stopped. The people saw two cannon facing them, pointing into their midst.

The man who carried the flag cried, "Friends! If I fall, remember my name is d'Arcole."

From the other side of the bridge came the command, "Fire!"

The man with the flag spun around and fell on his back, his head covered by the flag. Ten others lay on the bridge and the crowd fled, stumbling over the dead and wounded bodies of their comrades.

"The bloody bastards."

"They're firing grapeshot on the people."

"Fire on the gunners."

A commanding voice was heard:

"Stand your ground! Don't run."

It was the voice of Charras, a former student who had been expelled from the Polytechnical School five months before for singing the Marseillaise five months too soon.

Charras tried to step forward when he felt someone pulling his left hand. Looking down he saw a man kneeling at his feet gasping and trying to speak. Charras bent his head and saw blood running from the workman's chest.

"They got me. I'm dying. Take my musket."

He dropped Charras' hand and fell, knocking his head on the railing.

Charras took the musket, his face tense and calm as he fired. One of the gunners fell, clawing the cannon. Another shot was fired from the crowd and the second gunner fell.

An urchin said gaily to Charras, "Good work, citizen. Got any cartridges left?"

Charras looked at the dead workman. He answered mechanically, "No. No cartridges."

"You have a musket and no cartridges and I have cartridges and no musket. I'll trade you. I'll give you cartridges if you let me shoot. What do you say, citizen?"

Charras smiled and handed the musket to the boy.

Looking at the opposite side of the bridge, Charras saw that two new gunners had reloaded the cannon. He sprang back as the cannon was fired again. A bullet pierced the head of the boy, killing him before he had smelled the powder of the musket which he still held firmly. Many others were dead or wounded, and new gaps were made in the ranks of the assailants. Less than half of them remained alive, and among these only a few had weapons. The crowd wavered.

"Retreat, retreat."

"Forward to the City Hall."

They felt safer when they were together. But they were now packed into a dense group at the entrance to the bridge—an excellent target for the gunners who fired for the third time, covering the bridge with the dead. The soldiers closed their ranks, running with fixed bayonets to charge those who were still alive. The survivors dispersed in panic into the network of little streets buried in the heart of Paris.

Late in the evening two generals arrived at St. Cloud from Paris to see Charles X. They told the King that his crown was in danger and that he might still save it if he would revoke the ordinances.

The King listened graciously and, delicately wielding a toothpick, replied, "The Parisians are in a state of anarchy. Anarchy will necessarily bring them to my feet."

In this Charles X erred.

6 : The Glorious Thursday, July 29, 1830

Early in the morning M. Guigniault learned that during the night Galois had tried to climb the high wall which led to the rue du Cimetière Saint-Benoît. But he had not succeeded. The vigilant janitors had caught and brought him by force to the dormitory.

"What shall we do with Galois?" M. Guigniault was asked.

"Nothing," was his verdict.

M. Guigniault was performing his duty, and he was sure that he was performing it well. In the center of Paris he had succeeded in creating an isolated, peaceful island that remained neutral. The Preparatory School—and some day it would be the Normal School—had faithfully played the role that an educational institution ought to play: to work and study in isolation and seclusion. He had achieved what he wanted. His school had in no way influenced the struggle of the outside world.

Soon, however, he, M. Guigniault, would have to make a decision. He would have to declare himself for or against the Revolution. The decision must be well timed and well considered. Not only his future but the future of the Preparatory School depended upon it.

M. Guigniault was slightly disturbed. He realized that the Revolution might influence the fate of the Preparatory School. This influence could operate only in one direction: from the outside world to the inner world of the school. Was it quite right for the school not to influence the outer world? Of course it was right. But somehow he did not feel proud. He found himself thinking about Galois. He cursed this impertinent, strange boy, devoid of morals, of all respect and appreciation for the school. He clenched his fists and said to himself, "I must wait until this is all over; but I will show him, I will show him."

The hot sun shone on the streets which now stank of powder and blood. The Parisians looked at each other with pride and joy, for the fight was turning in their favor. In some places the people fraternized with the soldiers; in others the soldiers had been forced to withdraw. But they still held a firm line between the Louvre and the Champs Élysées.

Paris was crisscrossed with barricades. Students turned out in increasing numbers. Pupils of the Polytechnical School went through Faubourg St. Jacques, knocked at the door of every lodging house shouting, "Students! To the barricades!" The students' uniforms, especially that of the Polytechnical School, became signs of distinction. When Charras appeared near the Palais-Royal waving his two-cornered hat, over a hundred people surrounded him asking for leadership and ready for action.

"Where shall we go?"

"To the Prison de Montaigut."

Charras headed the procession. Behind him marched a drummer and a man carrying the tricolor flag. No fighting group was complete without these two.

The Prison de Montaigut was defended by a hundred and fifty well-armed soldiers. When Charras and his men arrived, they saw all the soldiers lined up in front of the prison wall, ready for the command of the captain who stood before them. Charras stopped his men. They spread out facing the soldiers, the opposing forces forming a simple geometrical picture: two parallel lines with two points between them. One of these parallel lines was formed by the soldiers, the other by the people. One of the points represented the captain and the other Charras.

The soldiers formed an orderly, straight line. Its uniformity gave the impression of overwhelming strength ready to be released by a single word of command. The line formed by the people wavered, was vague and disorderly. Some of the men were in rags, many of them thin and weak; less than half of them had muskets. Among them were children, a few students, and a very few well-dressed tradesmen symbolizing by their presence their assent to the people's revolution. It would seem that one order from the officer would make the undisciplined civilian crowd disperse in terror.

Charras was still far from the captain. He cried out, "I want to talk to you, Captain. May I come to you?"

"Yes, you may."

"Do you assure my safety?"

"Yes."

The soldiers relaxed and watched the scene. Charras approached the captain.

"You are a man of honor; you did not order your soldiers to fire upon us. I ask you in the name of the people to join their cause. Never was the cause of the people more honorable and more noble than today."

While Charras spoke, the pattern of soldiers and civilians began to change. The soldiers, instead of watching their adversaries, listened to Charras who spoke loudly in the tone of a man who knows how to stir emotion in simple hearts. The people, guided by correct strategic instinct and by a desire to hear what was said, moved slowly forward in steps that seemed accidental, aimless, and hardly noticeable. First one began to move, then his neighbor followed until this forward motion spread across the civilian line. Some of the men could now hear the captain's reply:

"I have sworn allegiance to the King and I shall not betray my oath."

Both the soldiers and civilians could hear Charras' well-modulated yet melodramatic voice:

"You swore an oath to the King. You feel bound by your oath because you are a gentleman and a man of honor. But did the King feel himself as bound by his oath as you do, sir? Did he not swear to respect the charter, and did he not betray his sacred oath?"

"I am not a politician. I am a soldier and I have my orders."

"If you refuse to join the people's cause, then at least remain neutral. Don't fire upon the people."

Then gesturing toward the civilian line he said, "Do not take their blood upon your conscience."

As he turned back to look at his men, he saw how near they stood to the soldiers. At a glance he understood his strategic advantage and how easily it could be increased, not by fighting, not by bullets, but by an uninterrupted torrent of words.

"We fight for freedom, freedom for France, and for the whole world. We want to bring back the glory of France that was hers after Marengo, Jena, and Austerlitz. We fight for a charter, we fight for the people. We want to return to the people their tricolor flag."

Charras saw the soldiers' faces hypnotized by the flag before them, and by the vision of Napoleon whom they beheld in this flag. He saw his own men, now only a few steps from the soldiers. He knew that the captain would not command his soldiers to fire, and even if he were to give the command, the order would not be obeyed. The captain looked relieved and smiled. He was glad that Charras had forced him into a situation in which he no longer had a choice. He stretched out his hand. The people and the soldiers fraternized among shouts of *"Vive la charte!"* "Long live our flag!" The soldiers offered their muskets, good army muskets, to the people. One more bloodless battle had been won.

The Duc de Raguse was at the Place du Carrousel, ready for a last and desperate stand. An officer brought news that at the Place Vendôme, soldiers had begun to fraternize with the people. The same old story! The duke decided to withdraw the rebellious regiment from the Place Vendôme and send the Swiss there instead. The Swiss in their red coats and bearskin caps were that day the only reliable defenders of the King. They did not speak French and had no brothers or sisters among the people. Soldiering, obeying orders, and shooting was their profession. Placed among those who hated them, they responded with hatred.

Two battalions of Swiss defended the Louvre. One was stationed at the windows of the picture gallery and in the colonnades, its colorful uniform an excellent target for snipers. But the Swiss returned the fire with skill and determination, repulsing all attempts to storm the Louvre. The second battalion of Swiss stood at ease in the courtyard, awaiting action. Meanwhile M. de Guise brought to the French commandant of the Swiss the duke's order to withdraw one battalion from the Louvre to the Place Vendôme. One battalion, thought the duke, would be sufficient for the defense of the Louvre.

The commandant who was to carry out this order decided to send the battalion defending the Louvre. These men were tired and needed a change. Therefore he would replace them with the reserve battalion in the courtyard. He intended to execute his plan in two steps. The first would be to let the defending battalion withdraw from the firing line, assemble in the court, and then march to the Place Vendôme. The second step would be to send the reserve battalion into the firing line.

The crowd outside the Louvre suddenly saw that the Swiss were withdrawing and that no red coats were visible. Without a command, without a preconceived plan, the people rushed into the Louvre. They broke the doors with axes; in a few seconds they scattered through the abandoned halls and fired through the windows at the red coats assembling in the courtyard. Musket shot followed musket shot. Amazement stirred the Swiss soldiers, changing quickly into fright and panic. They ran in disorder; they tumbled hurriedly over one another, making no attempt to return the fire that was decimating their ranks. They stampeded through the door leading to the Place du Carrousel, suffocating and trampling one another, flying in rout, and driven by fear.

The Duc de Raguse saw his last defenders in flight. He threw himself into their midst shouting, "Stop! Damn you, don't run! Form a line!"

But most of them did not understand French. Only fear guided their steps, and the fury which they had shown before in battle they now showed in flight. They crossed the Place du Carrousel, then the Tuileries, and scattered in all directions, casting down their weapons, tearing off their red coats, throwing them to the ground in the childish hope that it was the uniform and not themselves that the people despised and hated. The hurricane of their stampede was so swift that they drove before them cuirassiers, lancers, and police. They swept before them the remnants of the King's army which now fled in panic through the Champs Élysées.

The Louvre was taken! The Place Vendôme was taken! The Tuileries were taken! The tricolor flag waved over the King's palace.

The people hurried through the long galleries of the museum toward the palace of the Tuileries. The whole of Paris could now enter the magnificent apartments of the King and his family.

In the hall were statues of kings. The first wave of people had the opportunity of seeing them and smashing them; the second wave could only tramp over their splinters.

In the throne room a muscular worker, a large fellow with a cloth around his bleeding head, stood on the throne shouting, "I'm here in the bastard king's place." Then he jumped down and spat upon the throne. "Room! Make room on the throne!"

Four men carried a corpse in horizontal position and swung it violently. It was the corpse of a Swiss soldier, his red coat stained by red blood.

"He defended the King. Give him the throne as a reward."

They sat him on the throne and punched him under his chin in an attempt to make his head stay erect.

From the throne room the people pushed into the King's study. Here they rifled the drawers and threw the King's papers out of the window. Thousands of them fluttered into the Tuileries garden.

The biggest crowd was in the King's bedroom. People formed dense rings around the great bed of state which everyone wanted to glimpse. They rejoiced, laughed aloud, and commented upon the performance staged by two men coyly depicting the consecutive steps of love-making during which the passionate lover tore to shreds the duchess's gorgeous silver dress to reveal the rags of its wearer.

There was shooting in the picture gallery of the marshals. The favorite target was the portrait of the Duc de Raguse. One bullet pierced his head, two pierced his breast, the fourth missed and tore a hole in the background of the picture. A man then climbed onto the shoulders of his comrade, cut out the portrait in the shape of a medallion, and passed his bayonet through the duke's breast and head.

In the courtyard they danced a wild can-can to the piping of a fife and the scratching of a violin. Men wore plumed hats and the court dresses of the Duchesse d'Angoulême and the Duchesse de Berry. One of them wore a cashmere shawl over his rags. The can-can finished with a wild finale in which the shawl and all the dresses were torn to shreds. Their value was beyond the imagination of the people who wore them, yet all they wanted was to destroy the objects of luxury they saw.

When the troops were flying in disorder from the Louvre, a window was opened at the corner of the rue de Rivoli and the rue Saint-Florentin. From the other end of a sumptuous apartment came a broken and aged voice:

"Good God! What do you mean, Mr. Keyser, by opening the windows? You will have our house pillaged."

Mr. Keyser replied, "Never fear. The troops are in full retreat, but the people are interested only in pursuing them, not in plundering."

"Indeed," said Bishop Talleyrand, and he limped toward the clock. Then, in a solemn voice, he added, "Note it down, Mr. Keyser, that on July 29, 1830, at five minutes past noon, the elder branch of the Bourbons ceased to reign over France."

In the afternoon the students and teachers of the Preparatory School assembled in the Great Hall. They awaited the entrance of M. Guigniault and his announcement.

The door opened. Stiffly erect, the director made a triumphant entry, his face beaming, and a tricolor ribbon on his breast.

"Long live M. Guigniault!"

"Long live France!"

The director smiled, stretched out both his hands, quieting the waves of enthusiasm and devotion which now beat so strongly upon the rostrum. He started calmly, working himself up gradually according to the rules of oratory which he had learned so well and practiced now in a masterly fashion.

"Professors, colleagues, and students of the Preparatory School!

"This day, July 29, 1830, will live for a long time in the history of France and in the history of the whole civilized world. The tricolor, the flag of France, waves over Paris. It waves over the King's palace, it waves over the Louvre, it waves over Notre Dame, over all these places whose names are so dear to the heart of every Frenchman. We must not only wear these colors; we must cherish their image in our minds and nurse a love for them in our hearts."

Applause broke out, and when it died down the director began again solemnly.

"Professors, colleagues, and students of the Preparatory School! In the name of all of you, in the name of our Preparatory School, I declare our adherence to the provisional government of General Lafayette, General Gérard, and the Duc de Choiseul!"

The applause was wild. The same cries were repeated again and again:

"Long live General Lafayette!"

"Long live M. Guigniault!"

"Long live our flag!"

"Long live France!"

The director waited long and patiently until silence was restored.

"Let us now try to return to our normal school life. This is what all France must try to do and what we here must try to do. The examinations and the end of the school year are approaching. Let us hope that the government will again restore our Preparatory School to its rightful place as the Normal School of France, and revive its past prestige and importance."

This was the end. All that now remained for M. Guigniault was to wait until the cheers died down and to make a proper exit. As he looked at his students, his eyes were caught by a triangular face with eyes that looked at him as though he were transparent, and then they suddenly seemed to focus, wandering quickly from the top of his head to the tricolor ribbon, and saying with unmistakable clarity, "Men like you profane our flag."

The people had fought and died. With their blood and their bodies they had created a new chessboard. On this new chessboard, old hands were already playing the old game.

Fighting Paris was on the street. But the Paris of politicians, of many small, greedy men and a few noble and far-seeing men, was assembled at the palace of M. Laffitte. Here, in the house of this wealthy and ingratiating banker, was the center of intrigue: here the politicians devised schemes, here delegations were received, here the chamber had its permanent sessions; here, surrounded by thousands of spectators, was the Revolution's political brain and its political arm. The Revolution had no strategical center or military headquarters, but it had its political headquarters in Laffitte's palace. No, not the Revolution, but the bourgeoisie had their headquarters there; the same bourgeoisie who had incited the people to anger and indignation and who now in Laffitte's palace conspired for its own kingdom.

From Laffitte's house on this Thursday afternoon, Lafayette went to the City Hall to take over the command of Paris. General Lafayette was loved by the people, admired by the poor, and trusted by the honest. Around his head, the symbol of freedom and liberty, gleamed two halos of glory from two worlds and two revolutions.

The procession was greeted with joyous shouts:

"Make room for General Lafayette. The General is going to the City Hall. Hurrah for Lafayette!"

He had heard the same shouts forty-one years before! He had been crowned king of the people by liberty in 1789, and he was crowned again in 1830. The tired eyes of the old man spotted Etienne Arago, wearing a tricolor cockade. Lafayette turned to one of the men who accompanied him.

"M. Poque, go and beg this young man to take off his cockade."

Arago came to Lafayette.

"I ask your pardon, General, but I do not think I can have understood."

"My young friend, I beg you to take off that cockade."

"Why, General?"

"Because it is a little premature. France is in mourning. The flag ought to be black until France regains liberty. Later, later, we shall see."

"General! I have been wearing a tricolor in my buttonhole since yesterday and in my hat since this morning. There they are and there they shall remain."

"Obstinate fellow! Obstinate fellow," muttered the grieved old general on his way to the City Hall.

The City Hall again became the nerve center of Paris. Lafayette's room was full of people. Everyone wanted to pour out before the general the tale of his heroic deeds.

The general repeated to everyone, " Good, very good, excellent! You are a brave fellow." And he then shook hands with the men.

The recipient of this favor would rush down the stairs shouting to the people outside, "General Lafayette shook hands with me! Hurrah for General Lafayette!"

Charras, the ex-student of the Polytechnical School, arrived at the City Hall with his hundred and fifty men.

"Here I am, General."

"Ah! You, my young friend. I am happy to see you. You are indeed welcome." And the general embraced Charras.

"Yes, General, I am here, but I am not alone."

"Whom have you with you?"

"My hundred and fifty men."

"And what have they done?"

"They have acted like heroes, General! They took the Prison de

Montaigut, the barrack de l'Estrapade, and the one on the rue de Babylone. But now there is nothing left for them to take. Everything is taken. What shall I do with them?"

"Why, tell them to return quietly to their homes."

Charras laughed.

"Homes? You don't really mean that, General."

"I do, really; they must be very tired after the great work they have done."

"But, General, three-quarters of these brave men have no homes to go to, and the other quarter will not find in their homes either a piece of bread or a sou with which to buy it."

The general became sad.

"I should have thought of that. It alters the case. Let them have a hundred sous per head."

Charras went to his men and told them that the general wished to give five francs to each of them. It was a great sum for men in rags, but there was only one response:

"No! We don't want money. We didn't fight for money. Tell the general that we won't take a sou."

Charras felt like crying. His voice was tense with emotion when he delivered the last speech of a long day:

"Friends! You are the backbone and future of France, of the whole world. You are the great people of France. May our fatherland learn to know and love its true sons. Then it will become truly great."

The men looked at their leader with warmth and sympathy, understanding little of what he said.

"Friends! Let's celebrate our victory! You don't want money. But let me order bread, and meat, and wine, and here on the steps of the City Hall which we took today we shall eat our meal together."

"Hurrah for Charras!"

"Hurrah for Lafayette!"

To the deputies at his palace M. Laffitte said, "There is only one means by which monarchy can be saved—by crowning the Duc d'Orléans. The son of Philippe Égalité may appeal to the people's imagination. True, the son is not well known in France, but I consider this an advantage, because his strength will not come from the support of the mob. Therefore he will be forced to keep within the bounds where royalty should be confined. I have known and admired him for fifteen years. He shows his self-respect by admiring his wife; his children love and fear him. By placing him on the throne, we can save the principle of legitimacy in France and at the same time calm the revolutionary spirit of Paris. In the Duc d'Orléans we shall have a citizen-king."

The deputies knew that such a plan could only succeed if Lafayette supported it; he could calm the people or inflame them anew. Thus Lafayette must be watched, persuaded, and won over to the cause of the citizen-king.

A municipal commission of five was selected in Laffitte's palace. It was to form a ring around Lafayette to influence the old general and to cushion his contact with the people. Two bankers were among its members: Laffitte, the man of the day; and Casimir Perier, the strong man of the next two years. These five knew their task. They surrounded Lafayette with supporters of the Orléanist party. They sent the most ardent Republicans outside Paris, telling them, "Go and spread the Revolution over France." Among themselves they said, "Paris is now free of the most dangerous elements." They posted a sentinel at the door to Lafayette's office; he had orders to admit only members of a small camarilla. The noble old man was under the eyes of keepers; a prisoner in the house which he was supposed to rule. He was flattered and asked to sign unimportant documents and proclamations. He was the tool in a game of which he understood nothing.

But the people loved Lafayette. They believed that as long as the old general was in the City Hall, the future of France and the liberty and dignity of the common man were safe in his hands, and that no one could betray them.

In this the people of France erred.

7 : July 30, 1830

Galois left the school. Walking slowly along the rue St. Jacques toward the Seine, he looked at the bruised buildings, at the pavements from which stones had been torn, and at the remnants of a barricade.

"Here men fought and died while I made useless orations. Will I show the same courage when the test comes again in my lifetime?"

He wanted to escape from his thoughts and from his loneliness. He saw a small group of people surrounding a young man with curly dark hair and a sweaty face, who spoke with vivid gestures and repeatedly pointed to a placard.

The group was in a more or less stationary state, decreased by bored spectators, increased by idle passers-by. Galois joined the group and read the placard:

Charles X can never return to Paris; he has shed the blood of the people.

A Republic would expose us to horrible divisions; it would involve us in hostilities with Europe.

The Duc d'Orléans is a prince devoted to the cause of the Revolution.

The Duc d'Orléans has never fought against us.

The Duc d'Orléans was at Jemmapes.

The Duc d'Orléans will be a citizen-king.

The Duc d'Orléans has carried the tricolor flag under the enemy's fire.

The Duc d'Orléans alone can carry it again. We will have no other flag.

The Duc d'Orléans does not declare himself. He waits for the expression of our wishes. Let us proclaim those wishes and he will accept the charter as we have always understood and desired it. It is by the will of the French people that he will hold his crown.

Galois now listened to the quick, fluent voice.

"Here you see their greatest insult and perfidy. First they threaten you. They say that if you establish a republic it will mean two wars at once: a civil war and a war with Europe. Of course this is a lie. A republic would be so strong that no one would dare to attack it. And if they should, we could defend it. Who is in the army? The people!

The people are the soul of the Republic. They'll know how to defend it! But M. Thiers, who wrote this proclamation, thinks that you, like the Orléanists, are afraid of the rest of Europe. He ends by saying that you are the sovereign people, that you have the right to choose your own government. But you do not have the right to choose a republic, because it will be opposed, because there will be a civil and a foreign war at once. But you can still be the sovereign people of France if you go down on your knees and beg the Duc d'Orléans to accept the crown."

Here he paused, turned his back on the placard, and, facing his small audience, added excitedly, "Do they think we're so stupid? Yesterday we won the Revolution, and today, twenty-four hours later, we read a proclamation urging a new king. Did we fight Charles X to put the Duc d'Orléans in his place? We fought to uproot the Bourbon tree, not to replace one branch by another."

Galois liked the man who spoke and the way he spoke. He wished he could talk so as to be understood by everyone. But he disliked the attitude of the audience. They listened, they made comments, they agreed with the speaker, but they showed little enthusiasm. Where was the fire that must have burned yesterday? Where was the rage of the people that had just overthrown a kingdom and defeated an army?

"Let's tear down this placard so it won't deceive the people. Who will help me?"

Evariste thought, "I came here to be one of many, to learn how to talk to the people. I'll answer this young man even if I make a fool of myself."

He came toward the placard and said, "I'll help you."

Together they tore down the proclamation, and as they did so, Galois whispered, "You did a good job. Need any help?"

"Of course I do! Come with me."

This was a relief. The first contact was established much more easily than he had expected. They went in search of another placard, and the young man with the black curly hair said to Galois, "My name is Duchâtelet, I am a student at the Law School."

"My name is Galois. I am from the Preparatory School."

"Well! You're a rare bird. During the three days, we didn't see one student from your school. Glad to see at least one. How did you manage to get out?"

Galois blushed and started to answer, but Duchâtelet did not wait. He went on with increased speed.

"Pardon me for saying so, but your school is the most stinking school in Paris. You have quite a few sons-of-bitches there. We must get them!"

With difficulty Galois managed to put in a question.

"How do you know so much about it?"

"It is my business to know. We couldn't contact anyone in your school. It's good I found you. You may be very useful. We must create issues in your school to educate the students. They seem to need a hell of a lot of education. You'll be a splendid man for our first contact. What are you studying?"

"Mathematics."

"You must be a brainy fellow. Why aren't you in the Polytechnical School?"

Here was the question he feared. Always the same question! Someday he would answer it proudly and truthfully to the eternal shame of his two examiners.

He did not know what to say now, but Duchâtelet went ahead quickly and nervously.

"Of course you hate M. Guigniault's guts, your new tricolor director. I read in the paper today that he put his school at the disposal of the provisional government. He does not know that the provisional government doesn't exist."

Here, Duchâtelet burst out laughing, which allowed Galois to inject, "What do you mean?"

"I mean what I said. It just doesn't exist. During revolutions generals are made by tailors, and governments by printers. A journalist invented this government, announced it to Paris by proclamations, and here it is. Everyone swears that it exists. A good joke isn't it?"

He laughed again, and Galois asked, "Who is in power now?"

"There's Lafayette in the City Hall and a commission of five to see that he keeps his mouth shut. Four Orléanists with one Republican just for show. It looks bad. The people are sleepy and don't care much. But we shall awaken them! We must educate them, excite them, create issues until some day they will fight again. Why do I go on making speeches? We must go on with our work."

"Who are 'we'?"

"By 'we' I mean the 'Society of the Friends of the People.' It's the only active Republican society. It will gain strength now, you'll see. We have excellent men in it; M. Hubert, the lawyer, is our President. You must join the society, we need brainy fellows like you."

"How do you know I'm brainy?"

"Because you let me talk and you know how to listen and ask questions."

They saw another placard, and Duchâtelet asked Galois, "Would you like to perform in my place? I've done it five times already."

"I couldn't. But I would like to listen to you again."

Duchâtelet repeated his performance. Evariste was only half listening when the sound of his name awakened him:

". . . my friend Galois, a very fine mathematician, the most brilliant student of the Preparatory School. He can give you a good example of how those men behaved during the Revolution, the same men who are now for the Duc d'Orléans. Ask him about the director of his famous school."

Under the sudden compulsion to speak, Galois recited quickly as though it were a well-memorized lesson, "On Wednesday the director threatened us with the police if we went out and fought."

"Bastard," someone interrupted.

"And Thursday evening he wore a tricolor cockade."

"Bastard," someone repeated.

"You're right, citizens," said Duchâtelet. "This man and thousands of others think that we fought to exchange one Bourbon for another. They promise us the charter. But we can get the old charter at any moment from Charles X, who has already revoked the ordinances and promised to be a good boy. We want a charter, but not *the* charter. And look how cunningly M. Thiers speaks about the charter as though this were all we fought for."

They finished their performance by tearing up the placard and then went toward the City Hall. Even here the crowd was neither very big nor very much excited. Duchâtelet joined a few Republicans to whom he introduced Galois. They were alternately talking and listening to the various speeches delivered from the near-by stone post. At some indefinite moment Galois forgot to listen and was deep in his own thoughts considering the proof he had given in his paper to the Academy.

8 : July 31, 1830

Galois and Duchâtelet stood in the courtyard of the Palais-Royal. The crowd of well-dressed men and women shouted:

"The duke! We want to see the duke."

The cries were repeated until the duke and M. Laffitte appeared on the balcony.

"Long live the duke!"

"Hurrah for Laffitte!"

"Hurrah for the King of France!"

Then, tired by the repetition of these phrases, the crowd began to sing the Marseillaise. The Duc d'Orléans joined in a voice that was loud and out of tune.

Galois looked at the duke's broad jaw and narrow forehead, at his deceptively stupid-looking face, with its bushy sideburns. Only the small, deep-set eyes had a clever blink.

When the Marseillaise was finished, shouts began anew, then the

Marseillaise was repeated. The duke sweated and sang still louder and still more out of tune. Then he went down and mingled with the people in the courtyard. Among the bourgeoisie that surrounded him, the duke seemed like one of them. His gray hat, the black frock coat, the yellow gloves were like those of an ordinary well-to-do citizen. He carried around with great vitality his stout figure, his slightly extended stomach, and the fifty-seven years of his life. The future "citizen-king" went around shaking hands with everyone who put out his hand to him.

Duchâtelet whispered to Galois, "This man is dangerous! He knows how to become popular. Let's run away before he grabs our hands."

They went toward the City Hall, to the people who were ready to shout hurrah for Lafayette and who, thought Galois, would never shout hurrah for the Duc d'Orléans.

They saw a young man standing on a post and speaking to the crowd that surrounded him. Galois was fascinated more by the appearance than by the words of the orator. His suit looked as though it had been delivered by a tailor just an hour before. The waistcoat was white with silver buttons; the light gray frock coat had a perfect waist line, and his high hat with a narrow rim was silver silk. He looked almost like a dandy, out of place among men with dirty shirts and formless caps. He would have been booed and hissed except for the two big tricolor cockades, one on his hat and one on his coat. Because of these, people proudly listened to the well-dressed young man. It was hard to talk on this hot July day in the glaring afternoon sun, but the speaker looked cooler than anyone in the audience. Not a drop of sweat covered his large forehead, and his eyes were as penetrating as the two blades of sharp scissors. His face was handsome, manly, and cold. In his fingers he held a small bullet which from time to time he threw up in the air and then caught precisely at the point from which it had departed.

"Who is he?" asked Galois.

Duchâtelet answered, "Pécheux d'Herbinville. He is one of the members of our Society. Unlike most of them, he is rich and of an aristocratic family. I am sure he is proud of it."

Galois looked at the statue of self-control. His diction was perfect —a shade too perfect. When he emphasized words, he slightly shifted his lower lip to the right with an expression of affectation and cruelty.

The heat seemed less unbearable when one looked at this icy face and listened to the well-modulated voice:

"Who is the man whom they wish to give us as a king? I shall tell you, my friends, because being an historian, I studied the life story of Philippe Égalité and his son. The Duc d'Orléans is a bastardly descendant of Louis XIV. But he is a Bourbon and should share the fate of the Bourbons. They tell you now, M. Thiers and others, that he is the son of Philippe Égalité who voted for the death of Louis XVI. They tell you, the same gentlemen, that the duke fought for the Republic at Jemmapes. Squeeze any Orléanist and like a parrot he will repeat 'Jemmapes, Jemmapes.' "

Derisively, he squeaked the last two words twice, throwing the small bullet high into the air and catching it skillfully. Evariste, unlike others, did not seem amused.

"My friends, no Orléanist will tell you what I know and what you should know too.

"Thirty-one years ago, in 1799, the young Duc d'Orléans went to Mitteau, where the impotent Louis lived at that time. There, our Jemmapes hero threw himself at the feet of the fat Louis, wept bitterly, and cried, 'Forgive, my noble King, the crimes of my father and my own crimes, forgive me that I fought at Jemmapes.' "

He gave a funny imitation of the crying duke, but did not succeed in squeezing a laugh out of Evariste.

"And this man who kissed the infirm feet of a Bourbon king now wishes to become a king of France. Jemmapes indeed! There is only one man who can foil the intrigues of the Orléanists, and that is General Lafayette."

"Hurrah for General Lafayette!"

"I tell you, on excellent authority, that the duke has decided to win the general by honoring him today with a visit here at the City Hall."

"We don't want him at the City Hall."

"Keep him away from here."

A young boy came running and shouting, "They are coming! They are coming!"

Duchâtelet turned to Pécheux d'Herbinville, who descended from the post and stood right before him.

"You made a very good speech, but I am afraid it will be of little use now."

"Yes, I am afraid so."

"This is Galois, a mathematics student at the Preparatory School. He is with us."

They shook hands, and Pécheux said in a slightly patronizing tone, "We can use a student from the Preparatory School."

He tried to smile but he didn't quite succeed.

The procession arrived. The Duc d'Orléans rode at its head on horseback; he was dressed in a general's uniform and wore a large tricolor cockade on his hat. He looked straight ahead as though hypnotized by the steps leading to the City Hall. The banker Laffitte, in a sedan chair, was carried by Savoyards; he had sprained his ankle and was in pain. The duke and Laffitte were followed by eighty deputies. It was a poor spectacle for those who had witnessed the splendor with which Charles X appeared on the streets of Paris.

The people in the square showed neither hostility nor approval; they were silent as the duke slowly neared the steps. His white horse dispersed the crowd, which gave way calmly and frigidly. The duke's face was now ghostly pale. When he reached the City Hall, he stepped down from the white horse and with a firm tread began to climb the steps. At this moment, General Lafayette came out and waited at the top of the stairs on which the duke mounted higher and higher, slowly approaching the level at which Lafayette stood. He must reach this level if he was to push the old general down so that he could never climb up again. Was Lafayette the only man who did not understand that?

The general greeted the duke with the politeness of a gentleman who knows how to treat a distinguished visitor. Then the whole procession disappeared into the City Hall.

All eyes were now directed toward the front of the City Hall. Everyone expected something to happen, and this expectation slowed the flow of time.

Duchâtelet turned toward Galois and asked, "Do you think that the general will stand up to the duke?"

"No."

Duchâtelet asked Pécheux d'Herbinville the same question.

Pécheux answered, "I don't know." And then looking down at Evariste said, "Why are you so sure that he won't?"

"Because I know history."

Pécheux answered tartly, "Indeed. For a mathematician you seem to be quite an historian."

"Your remark. . . ."

Duchâtelet interrupted Galois:

"They're coming."

Lafayette led the Duc d'Orléans onto the balcony of the City Hall. The two men looked silently at the silent crowd. Then Georges Lafayette, the general's son, handed his father a folded tricolor flag. The old general began to unfold it. At this moment, for the first time on that hot July day, a cool breeze blew gently. It put life into the tricolor. The flag fluttered away from the general's shaky hands and covered the duke's face. The duke took the flag in his thick fingers while Lafayette turned the pole. The people saw the tricolor increasing in size, firmly held by the two men. They saw neither the shaky dried-up hands of the old general, nor the duke's fingers greedily digging into the pennant.

The surface of the flag rippled toward the people and away from them. Suddenly the icy silence was broken by loud cries:

"Long live our flag!"
"Long live Lafayette!"
"Long live the Duc d'Orléans!"
Galois turned to Duchâtelet.
"Let's shout, 'Down with the duke!' "
Pécheux said, "I disapprove of useless demonstrations."
Their words were drowned by cries:
"Long live Lafayette!"
"Long live the Duc d'Orléans!"
It was the dying cry of the Revolution. The people's part was ended; the reign of the bourgeoisie had begun.

9 : 1830

In August, Charles X, King of France, was exiled and Louis Philippe was proclaimed King of the French. The King of France was replaced by the King of the French; the older branch of Bourbons by the younger branch of Bourbons; the reign of the aristocracy by the reign of the bourgeoisie.

What did the people gain who had fought and died for the charter and the flag? First with astonishment, later with anger and hatred, they found their sufferings increased by the Revolution. They had hoped that now work would be easier, bread more plentiful, their children better dressed and better fed. They had hoped that the July days would relieve their misery. But they won nothing.

Before the Revolution, the biggest printing office in Paris employed two hundred workers, each of them earning about five francs a day. When the Revolution broke out, the printing offices closed. After their reopening, ten men were taken back, and during the next six months the number of workers increased to twenty-five, at wages of two francs a day.

Some citizens suggested that the Minister of the Interior should found a large printing press that would be owned by the state and would reprint the revolutionary works of Rousseau, Voltaire, and the

Encyclopedists. This, they argued, would raise the level of education and diminish unemployment.

The Minister of the Interior rejected the proposals and stated his reasons: "Such books will not find a sale. They are old weapons, useless for liberalism now, after its battle is won."

What did the government do as capital disappeared, bankruptcies spread over Paris, and the economic depression widened? The government roused the people's anger against the Republicans by repeating the same arguments in its press and proclamations hundreds of times:

"You, the people who won the Revolution, are the backbone of France. You fought and you achieved all that you strove for. Do not be deceived by the Republicans. They want you to fight again under their leadership. What will they do if they win? They will drive you into utter poverty! They will declare wars on Europe! They will not rest until France's soil is invaded and your misery a thousandfold increased!"

The manifestos of the Republicans were indignantly torn down and their authors branded as men who thirsted for pillage. Once a crowd of men invaded the offices of the Republican paper *Tribune*. They were in rags, poor, and dirty; not one of them wore a black coat or yellow gloves. They broke into the offices, smashed the printing press, destroyed the furniture, and shouted, "Away with the Republicans. Shoot the Republicans. Kill the bloody bastards!"

Lafayette, still the commander of the National Guard, sent a detachment in time to save the lives of editors and writers.

The people oscillated in their wrath between the Orléanists and the Republicans. They knew that they were deceived, but they did not know by whom. The seeds of a new revolution were sown the very day the July revolution ended.

One of the first acts of the government was to change the Preparatory Schools into the Normal School and the two-year into three-year curriculum.

Galois passed the yearly examinations and was promoted into the next class. Free from school, he spent the vacation in Paris and joined the Society of the Friends of the People, at that time the most influential Republican organization. In it he belonged to a cell of students who planned events in the schools, wrote pamphlets, organized lectures, and excited by long discussions their own hope for a new revolution and their hatred of Louis Philippe.

Here Galois made new friends and new enemies. He learned that the most secret plans of the society became known to the police, that the society was full of spies. Often he was afraid to sound too radical, because he knew that this was a characteristic of provocateurs, who pushed the society to adopt the most violent measures, to create riots of which the police knew in advance and against which it was well prepared. He found out that even among the Republicans were not only men worthy of love and admiration, but also men whom one had to despise and fight; here he found heroism and cowardice, honesty and crime, brilliance of mind and plain stupidity; an atmosphere sometimes depressing and sometimes exhilarating.

Auguste Chevalier, who had just graduated from the Preparatory School, was the only non-Republican whom Evariste saw often. Chevalier was Galois' only contact with the scientific world, the only man who always listened admiringly to his friend, trying more persistently than successfully to understand Evariste's mathematical work.

Once Auguste asked Evariste what had happened to the manuscript he had sent to the Academy more than half a year ago, in February. When Galois replied that he had never received an answer, Chevalier convinced his friend that he ought to go to the Institute to find out and even insist on seeing Professor Arago if necessary. Galois promised.

Two days later, when Chevalier came to Galois' room, his first question was, "Well, Evariste, did you go?"

"Yes."

"And what did they tell you?"

"Nothing."

"What do you mean?"

"They lost my manuscript."

"I'm not asking you about your first manuscript. What happened to your second manuscript?"

"They lost my second manuscript."

Chevalier seemed more depressed and unhappy than Galois. He said, "Please, Evariste! Tell me what happened as fully as you can."

"There's not much to say. As you know, Fourier was the Academy's secretary and he died recently. Whether he kept the manuscript or sent it to someone else, no one knows. It was not found among his papers. There's a possibility that the manuscript went to M. Cauchy. Fortunately no one suggested that I see him, because M. Cauchy is in exile. He is a pious man. He may lose a manuscript, but he will not betray his oath to Charles X. On the other hand, Charles X is a great king. He will know how to appreciate devotion and he will certainly reward M. Cauchy."

"Please, stop. I feel like crying."

"How do you think I feel?"

"What else did they tell you?"

"I saw M. Arago and M. Poisson. They were both very polite. M. Poisson insisted that I rewrite my paper once more and send it to the Academy. This time, he said, he will take special care that it won't get lost."

"You ought to do it. You must do it!"

"It's not as simple as you think. I wrote the manuscript over half a year ago. Now it seems antiquated to me. I'm still working, but I have little patience with writing for fools who don't want to understand."

"Perhaps they will, perhaps someday they will!"

Galois repeated mechanically, "Perhaps they will, perhaps someday they will."

The sounds of these words seemed to have excited him for he began to speak louder and louder, almost shouting, "No doubt they will. Surely they will understand. But when? Oh yes, someday they will understand." Then he burst out with a short laugh. "Don't worry, Auguste. I shall write it down before I die. And if I cannot print my papers, then I shall make you the custodian of my manuscripts."

He still laughed. Chevalier said meekly, "No, Evariste. It isn't funny. I don't like your brand of humor."

10 : September 17, 1830

The Society of the Friends of the People held a public meeting every week in Pellier's riding school on the rue de Montmartre. The members of the society sat in the enclosure of the riding ground, separated by a wooden balustrade from the extensive corridors which were left open to the public. The meeting on September 17 was well attended. The president, M. Hubert, sat at a table, and at his right was Godefroy Cavaignac, one of the most admired orators, one of the most beloved Republican leaders. His graceful tall figure with its military bearing, his heavy mustache, his firm eye with a shade of sadness, made him look like a girl's dream picture of a hero.

On the left side of the president sat Raspail, blond, small, young, in his thirties, a writer who knew how to use irony and a scientist who knew how to use imagination. He was one of the few famous scholars who unreservedly allied themselves with the people and their cause.

M. Hubert introduced the subject for discussion: What should be done with the four imprisoned ministers of Charles X? Did they deserve death? And if they did, then how could the intrigues of the King, his ministers, and all the deputies who wished to save the lives of these enemies of the people, be prevented? When he finished his introductory remarks, he gave the floor to Godefroy Cavaignac.

The speaker began calmly, without the oratorical effects which, as everyone knew, would come in due time. His voice had warmth and fire. Every word was heard by everyone. He described the crimes of the ministers, their wickedness, their stupidity, their plight and imprisonment.

Then he asked, "Why do the King and the ministers preach clemency, instead of justice? Why did M. Victor de Tracy propose in the chamber the abolition of the death penalty? If these little men who now govern France had answered this question honestly, they would have told you, 'France is a small, weak country, afraid of every other country in Europe.' They would have told you, 'The death sentence to the ministers will not be liked in England and Russia, and the times when France did not fear England and Russia are gone.' This is what they would have told you if they were not cowards and hypocrites. They want to remove the death penalty and they wish to prepare your minds for clemency. Once this is done the rest will be easy. They will entrust the passing of sentence to the peers of France, many of them devoted friends of the ministers. It is they who will judge the July murderers."

He raised his voice.

"So this is what they intend to do! The scaffold for obscure culprits, and for illustrious criminals—impunity! Let a tormented man commit murder in rage or out of despair and who would care about saving his head from the knife of the guillotine? Everyone would be ashamed to bestow compassion on this crime, even if its source were despair, poverty, and unhappiness. But let nobles, let rich men, men who hold in their hands the destinies of empires, sacrifice a million human beings to their pride, set a city in flames, force brothers to cut each other's throats, their families to suffer forever, let them do all that and more; and then when the hour for revenge comes, you will hear of nothing but clemency, the glory of pardoning will be cried up, and law will at once relax its rigor.

"They tell you that they want to keep the Revolution pure, to let it shine with generosity as it has shone with the noble light of courage. Well, then, let the task of judging Charles X's ministers be confided not to the Chamber of Peers where they have their relatives, their friends, their allies, and their accomplices. Let the task of judging the ministers be confided to a national jury especially enrolled for that grave office. And let the jury condemn them! Let it condemn them to death! Because if they do not deserve death, if they do not deserve ultimate punishment, then they deserve none. Then when that sentence, that sentence of death shall have been pronounced, let an appeal be made to the clemency of the people. Let the people exercise the right of grace and clemency. It showed itself great enough, God knows, when, with the absolute mastery of Paris, it knew how to keep itself within bounds, and the properties of the rich were protected by men who use the church steps or the stones of the streets for their beds.

"But no! The generosity of the people which they extol in idle, bombastic words is calumnied, regarded with fear and dislike. They are afraid that the people may make too glorious a use of its victory. They are afraid that in granting clemency, the people will manifest its virtues as before it manifested both its strength and its virtues. If it is for the sake of revolution that they wish to pardon the ministers, then let them not address themselves to those who passively received and later betrayed the Revolution. Let them address themselves to those who made the Revolution, who gave their blood in the three glorious July days. Let them address themselves to the people of France!"

It was a long time before the applause died down. M. Hubert looked around to see who wished to speak next. He saw Galois' raised hand. This was to be Evariste's first public speech. He would have liked to cram what he had to say into a few sentences, dryly formulating his thesis and reasoning. But he had learned by now that arguments will be listened to and accepted emotionally only if surrounded by oratory, by words and words, often unnecessary, often even meaningless, but with the magic power of evoking passion. He had learned that a Republican must know how to stir anger and pity, hatred and love.

When Galois rose, he saw the faces of people darkened and distorted as though a dense curtain of fog covered his eyes. His own words, spoken loudly, seemed strange in their sound, and he detected hesitancy—the unmistakable stamp of fright.

"Citizens! Our problem is only a special part of a much wider problem, which is: Has the state the right to dispose of human life?"

There was cold indifference to this opening sentence. Galois wished someone would remove him or the audience by force. The burden of continuing the speech seemed unbearable. He collected all his courage to recite the next sentences and to say them properly.

"This is the question Louis Philippe and his ministers ask now, when they have to decide the fate of Charles X's four ministers, the men whose hands are stained with the blood of the people."

Galois felt the bad taste of the last few words. They sounded cheap and trivial. But they had an effect. There was even some weak applause. Its noise thinned the curtain of fog before his eyes.

"There are among us those who think that the people ought to show their generosity and not demand blood for blood and life for life. Let us assume that we do not take the lives of these ministers but imprison them instead, for one, two, or even five years. During this time passions will die away, public and private sorrows will be appeased. Someone will start a new cry for clemency, someone will again appeal to the generosity of the people. The history of our struggles, engraved with musket balls and grapeshot on the walls of our town, will no longer be legible. Then a voice will demand that the ministers be freed and banished from our country.

"They will leave France. They will go to foreign lands and they will intrigue with foreign powers against the people of France, whom they have always hated and despised because they did not give in to their lust for power. And the same men for whom clemency is now asked may come back to France as victors, after thousands of Frenchmen are killed. They may come back as victors to subdue our country and to increase the misery of its people. Or, perhaps, they will be allowed to come back as free men and try once more to renew their greedy grasp and extract from the people a price in blood and lost liberties as a reward for its generosity."

Galois now felt that he was listened to. The curtain of fog had lifted. He no longer needed to use his memorized sentences. He felt the joy of speaking to the people and of being heard by the people.

"I ask you, is the assumption that I just made fantastic? Is it not exactly what happened to France before? Has not the expelled aristocracy always allied itself with the enemies of the French people? They care only for power, property, and titles. They did not care and they will not care for the people. These men learn nothing and forget nothing."

It was his father who had said to him, "The Bourbons learned nothing and forgot nothing." The thought of his father magnified his emotion and increased his will to convey it to the people. His voice grew stronger.

"By taking four lives now we may save thousands, perhaps hundreds of thousands of lives later. We must decide whether we are for the people or, be it by stupidity or wickedness, against the people."

He felt that he was losing control of his emotions, that a voice stronger than his own was now speaking through him.

"When the incorruptible Robespierre asked for the head of Louis Capet he said to the convention, 'The King is not an accused man, you are not judges. You are, and you can only be, statesmen and representatives of the nation. You are not called upon to give a verdict for, or against, a man; but to make a decision for the public weal, to exercise an act of providence for the nation.' Allow me to say to you today, the ministers must die so that the people can live in peace and safety. There is only one slogan for us: *Death to the ministers!*"

The public outside the balustrade responded with applause and choral repetitions: "Death to the ministers!"

Not all the members applauded. Some of them looked at Galois with a mixture of anger and astonishment. Evariste saw the face of Pécheux d'Herbinville, who smiled ironically, whispering to his neighbor; and he caught a glimpse of Raspail, who nodded sympathetically. Galois did not listen to the next speakers, but turned endlessly in his mind the sentences he had said and those he ought to have said but had forgotten. From the small fragments of speeches that penetrated his mind, he understood that not all the Republicans desired the death of the ministers. But the public interrupted many times with shouts that he heard with joy:

"Death to the ministers!"

When the meeting ended, Raspail came to Galois and said, "I liked your speech, Galois."

Evariste blushed and answered, "I'm glad you did."

They went out together along the rue de Richelieu, then turned to the left, around the Louvre to the Quai de l'École.

Raspail interrupted the silence:

"I liked your speech for its logic and precision. You said exactly what I wanted to say. You seem to know that the most important thing is to make the people understand the issues, to make them conscious of what happens, and to strengthen their will to fight." He talked more to himself than to Galois. "We can't do anything without the people. When will they burst out again in anger and a revolution which will sweep away Louis Philippe's throne? Is this issue of punishment for the ministers important enough to cause a successful revolution? No one knows."

Galois asked in a whisper, "Do you believe that the people may rise soon?"

Raspail looked dreamily at the Seine.

"Who knows? We are still weak. We are divided among ourselves. We seldom agree on tactics and issues. There are among us men with whom it is difficult to work, spies who appear to be the most ardent Republicans. We have to fight not only the regime of Louis Philippe, but the Bonapartists who would like to see Napoleon II on the throne of France, the legitimists who would like to see Henry V on the throne of France. But our time will come. If not in mine, then perhaps in your lifetime. I believe that by our efforts we are bringing that day nearer. Don't you believe in this?"

Galois said so softly that his words were hardly heard, "I do believe in it."

There was silence between them. Then they heard a noisy crowd turning toward the Quai de l'École. Soon its shouts became distinct:
"Death to the ministers!"

11 : December, 1830

When Evariste entered the second year of the new Normal School, he knew why he was there. It was not to learn, but to stir up trouble; not to plow through two more unbearable years, but to spread love of the Republic and mistrust of the director. Yet all Evariste achieved was to change the unfriendly indifference of his classmates into hatred. The campaign at the Normal School planned by Evariste and his Republican group was a failure. What it did was to cover Galois with ridicule and M. Guigniault with added prestige. A spectacular stroke was planned for December.

After Evariste struck, strange things began to happen. The masters ceased to look at Galois or ask him questions. His comrades glanced sideways at him, whispered among themselves, and froze in silence when Evariste came near. Students were called to the director's and masters' studies, returning with an air of importance and mystery. Then the electrically-laden atmosphere produced a sudden storm.

On December 9, M. Guigniault came into the study room where all the pupils of the Normal School were assembled. He was flanked by M. Jumel, the subdirector, and M. Haiber, the master. In his hand he shakingly carried a newspaper. He began to speak in a voice that now vibrated with sorrow rather than anger. There was no place today for flashy oratory, and only the experienced listener could detect how artfully the tones of pain were swelled and those of anger suppressed.

"I must talk to you about a very serious matter. There is a Judas among you!"

The chill was successfully produced. Now it must deepen until the well-timed explosion came.

"I have before me a rag that calls itself *Gazette des Écoles.* In it is an article in which I am unjustly abused in filthy, vulgar language. I am sure that those of you to whom I am addressing these words will believe me when I say that the appearance of this article means nothing to me. If it were only the article I would have thrown out the filthy rag and forgotten quickly all about it.

"Something unbelievable and monstrous has happened! But I cannot tell you without reading a piece of filth. Forgive me, that in doing so I shall soil the air of our Normal School."

He put on his glasses.

"I shall read the end of the article, because, as you shall see, I *must* read it to you, and I beg you to listen to every word. The so-called editor writes at the end:

" 'We cannot better continue our article than by quoting the following letter that we have received:

Gentlemen:

The letter which M. Guigniault has today inserted in the *Lycée* about one of the articles in your paper has seemed to me very inappropriate. I have thought that you would eagerly welcome every means to unmask this man.

Here are the facts which can be confirmed by forty-six students.

On the morning of July 28, many of the Normal School students wished to leave the school and fight. M. Guigniault told them on two occasions that he could call the police to reestablish order in the school. Indeed, the police on July 28!

On the same day, M. Guigniault told us with his usual pedantry, "There are many brave men fighting on both sides. If I were a soldier I would not know how to decide—whether to sacrifice liberty, or my oath to the King."

Such is the man who next day covered his hat with a tricolor cockade!

Everything in him stands for the narrowest ideas and most complete routine. I hope that you will be glad to receive my information and that your admirable paper will make as much use of it as possible.' "

He put the sheet on the desk.

"I would not have bothered with the letter, the falsehoods and slanders it contains, if not for the signature. The letter is signed, 'A student of the Normal School.' It also contains an editor's note which I shall read to you too:

> In publishing this letter we decided to withhold the signature, although we were not asked to do so. Let it be noted that immediately after the three glorious days, M. Guigniault announced in all the newspapers that he had put all the students at the disposal of the provisional government.

He took off his glasses, played with them and said, "It seems unbelievable that there could be among you even one who would do this evil, cowardly thing to our Normal School, to his school comrades, and to me. There is only one thing for me to do, that is to find out, to ask each of you separately."

With a commanding gesture he turned toward the first student in the first row.

"Did you write this letter?"

"Certainly not, sir."

Toward the second student: "Did you?"

"No, sir."

Eight questions more, he counted, then he would have his man in his clutches. It would take only one minute more.

To the third: "Did you write the letter?"

"No, sir."

"Did you write the letter?"

There was no answer. Something had gone wrong in the preconceived plan.

He repeated sternly, "I ask you for the second time, did you write the letter?"

His finger pointed to the fourth student in the first row.

Finally the answer came:

"Sir, I do not know how to answer that question."

"You don't know whether you wrote the letter or not?"

"Yes, sir, I know. But I know, too, that by answering your question with 'no' I am denouncing one of my comrades."

He thought, "The bastardly young puppy from the first year! How does he dare?" Without any transition, M. Guigniault's calmness collapsed suddenly and completely. He banged his fist on the desk and shouted violently, "You are afraid to denounce you comrade! What nobility! What loyalty to Judas! Today, young man, you are shielding a crime; tomorrow you will be its accessory." His shouts burst into a raging flood. "I know who did it, I know the burning sore on the flesh of our school, the Judas in our midst!"

He stepped from the rostrum to the second row, pointed to Evariste and cried, "You! You did it! I defy you to deny it!"

"Sir."

"Don't dare to speak. Never let me see you again. Never! Did you hear me? Never! Run and pack your things. We shall get rid of the greatest troublemaker this school ever had. Out with you!"

"Sir! You don't have the right. . . ."

"Keep quiet or by God I shall lay my hands upon you. Out with you! I don't want to see you and I don't want to hear from you. Out with you. . . ." He turned toward the two men on the rostrum. "M. Haiber and M. Jumel will take care of you. Today you go, and never again do I wish to see you or hear your name."

He banged the door, went to his study and sank into a chair, wiping sweat from his forehead. He cursed Galois and he cursed the fourth student in the first row. He took a piece of paper and began to write the draft of a letter to the Minister of Education. His uneven, shaky handwriting became neater and calmer; the more self-righteous was his argument, the more abuse he poured upon Galois' head. He wrote:

Dear Sir:

It is my very painful duty to give you an account of an act for which I had to take full responsibility and for which I ask immediate ratification.

I have just expelled the student Galois from the Normal School and had him taken home to his mother, for reasons indicated in the letter which I had the honor to write to you the day before yesterday. This student's action aroused the indignation of the whole school. It concerns a letter inserted in the issue of that very day of the *Gazette des Écoles*—since it must be called by its name—and signed, "A Student of the Normal School." To everyone who read and also spoke to me about it, this letter seems to compromise the very honor of the school so seriously, indeed, that it was impossible for me to ignore it.

Since all indications pointed to Galois as the author of the letter, I thought that I had no right to leave the whole school under the weight of one man's guilt and that once the culprit was recognized, he and I could not remain together under the same roof. I, therefore, expelled him at my own risk, and in this I only did tardily what I had been tempted to do twenty times in the course of last year and ever since the beginning of this.

Galois is, in fact, the only student against whom the professors and the tutors have complained continuously since his entry into the school. But I was prejudiced by the idea of his incontestable talent for mathematics and I mistrusted my own impressions, since I had already had reason for being personally dissatisfied with him. Therefore, I bore with the irregularity of his bahavior, his laziness, his unruly disposition, not in the hope of changing his character, but in the hope of guiding him to the end of his two years without bringing sorrow to a mother who, I know, counted on the future of her son. All my efforts were in vain and I recognized that the evil was without remedy; there no longer exists any moral feeling in this young man and perhaps there has been none for a long time.

He looked at the last sentence. He mumbled to himself, "There no longer exists any moral feeling in this young man." He repeated mechanically the same words. They soothed his anger and restored his self-respect.

The director waited impatiently for the minister's reply. It took almost a month for the rusty bureaucratic wheels to throw the expected letter into M. Guigniault's hand. He read it with relief.

"So the minister approved! So this is the end of Evariste Galois so far as I'm concerned! I'll never see him again! I'll never hear about him again!"

But hear he did. First it was painless, almost pleasant, while glancing through the daily papers to read about the wicked Galois. It was good to know that there it was, printed for all France to find out, what he had known some time ago: that there no longer existed any moral feeling in this young man. With pride he informed his colleagues:

"I knew him, he was always a good-for-nothing. I had to expel him from the Normal School."

Around 1850 Evariste Galois became known as a mathematician. M. Guigniault was then fifty-six years old. When asked about his former student, he used to say, "Galois as a young man showed genius for mathematics. We, in the Normal School, always knew it—very much unlike the foolish examiners in the Polytechnical School who flunked him twice. Can you imagine such stupidity?"

"Did he finish the Normal School?"

"No! As far as I remember he knew too much mathematics and left our school after the first year."

In 1870 a famous French mathematician, Camille Jordan, wrote a long (667 pages) book on the theory of substitutions. He said in the preface that his work was only a commentary to Galois' papers. It was this book that made Galois' theory known over the entire mathematical world and his name shine with increasing glory, until it became one among those few most illustrious and famous that will be recorded forever in the history of mathematics.

In 1876 M. Guigniault was eighty-two years old, and this was the last year of his life. He had been asked many times by mathematicians and nonmathematicians about his former student, Evariste Galois. By that time M. Guigniault was tired and apathetic. He had lived too long and had seen too much. He remembered three revolutions, the reign of three kings and two emperors. Through a toothless mouth he repeatedly mumbled the same answer:

"Galois! I remember him. He was a strange boy, a very strange boy."

VI

"TO LOUIS PHILIPPE"

1 : Tuesday, December 21, 1830

AFTER HIS expulsion from the Normal School, Galois joined the third battery of the National Artillery Guard. He bought the colorful and expensive uniform: a blue military coat with red epaulettes, a forage cap with a red ball of horsehair in front, and trousers striped with red. Twice a week he drilled in the quadrangle of the Louvre between six and ten in the morning, and once a week he practiced shooting at Vincennes.

The National Guard was the sword of the bourgeoisie. Nominally it was open to everyone, but a simple device served to eliminate the poor: all members had to buy the expensive guard uniforms, and men in rags had no money for uniforms. Among the members of the National Guard, the Republicans were spread too thinly to change its character or influence its actions.

The new slogan of the Republicans was "Join the National Artillery Guard." Four batteries comprised the artillery. The second and third battery had a Republican majority. Perhaps half of the members of the fourth were Republicans too. They were in a minority only in the first, of which Louis Philippe's son was a member.

It was December 21, the day of hope for the Republicans and the day of fear for the government. The Republicans were prepared. Their Artillery Guard was prepared. But Louis Philippe was prepared too. And so was his army and his National Guard.

It was the last and decisive day of the trial of Charles X's ministers. The House of Peers in the Luxembourg would soon proclaim the verdict. Troops and national guardsmen blocked all the streets around the Palais des Pairs. Two squadrons of lancers and six hundred soldiers stood at the south gate leading from the observatory to the garden. Altogether, thirty thousand uniformed men encircled the palace. Around them and among them was the dense and restless crowd of Paris.

The people cried:

"Death to the ministers!"

"To the Luxembourg!"

"Death to the ministers!"

Less than half a year before, the cry *"Vive la charte!"* had passed from the bourgeoisie to the people. Now the slogan "Death to the ministers," coined by the people, confused the National Guard, the defenders of public order and private property. They remembered the nation's unity in the July days. Some of them would have joined the people as the people had joined them five months before. But there was one thought that broke this unity: "The people will pillage if you do not keep order."

The National Guard stood firm.

On the Place du Panthéon, Professor Arago met a group of men armed with clubs and repeating the cry of the day:

"Death to the ministers!"

He warned them that they were playing into the hands of their enemies, that they were furnishing an excuse for the use of brutal force—a force that would be turned against them. The lofty speech of the great scientist and liberal was interrupted.

"Shut up! We don't want to listen to you."

Arago became excited.

"Don't you understand that I share your views?"

"Men with coats of different cloth can't have the same views."

The man who said this clenched his fist around Arago's jacket and pushed him against a lamppost. At that moment, a loud cannon shot sounded.

"To arms! To arms! To the Louvre!"

And the crowd, leaving Arago at the lamppost, rushed in the direction of the Louvre.

At the same time, at the Place de l'Odéon, Lafayette spoke to a crowd. The old man expected the same reverence and enthusiasm that had greeted him always and everywhere. But today the crowd was angry. It shouted into his face again and again:

"Death to the ministers!"

Lafayette talked to them as to children who misbehave.

"Go to your homes! I ask you to disperse peacefully."

No one moved.

"I do not recognize here the combatants of July."

A man replied, "That's very likely; you weren't there."

At this moment a cannon shot sounded.

"To arms! To arms! To the Louvre!"

And the crowd left Lafayette and rushed in the direction of the Louvre.

When Louis Philippe heard the cannon shot, he sighed with relief. He understood its meaning. This was the signal that the prisoners had arrived safely in Vincennes. He knew that they would not be condemned to death, but he had feared what might happen on their journey to the prison. And as to the rest? He was prepared! He would not lose the battle as Charles X had done, by stupidity and weakness.

On December 21, Galois, with all the other artillerymen, was stationed in the Louvre quadrangle. The plan of the Republican members of the Artillery Guard was simple. Indeed there was but one thing wrong with the plan: it was too simple. It forgot that the new regime was cleverer, more ruthless than the stupid old regime of Charles X.

This was the plan: The Artillery Guard was to stay in the Louvre during the last day of the trial. In the July days the climax of the

Revolution had been the defeat of the Swiss and the storming of the Louvre. Now, however, even before the fight began, the Louvre would be in the hands of the artillerymen, who—in the majority—identified their cause with that of the people. Here they would await the people, then they would open the gates, deliver the cannon to them, and join their fight.

In the guardroom of the Louvre the artillerymen discussed the events of the day, literature, science, sex, and played cards. An artilleryman came in and whispered to Bastide.

The senior captain of the third battery exclaimed excitedly, "Impossible!"

The artilleryman said, "See for yourself."

Bastide commanded, "Men of the third battery, come with me."

They took their muskets and ran into the courtyard. There they saw a group of artillerymen from the first battery dismantling the cannon. Bastide sprang into the middle of this group with his sword drawn.

"Out of here! Out of here instantly or I swear I will put my sword through every one of you."

An officer stood up and said, "Captain Bastide! I am Commandant Barré. . . ."

"I don't care if you are the devil himself. Out of here! My orders are that no one is to touch the cannon without my permission, so out you go."

Barré and his men withdrew. The artillerymen prepared the cannon, and Bastide left a sentry in the courtyard to be changed every hour. Galois and Duchâtelet, both members of the third battery, volunteered for the first watch. When they were left alone, Galois said:

"Nothing will come of this. We shall fight here among ourselves instead of fighting together with the people. Just because we once won a revolution without expecting it, just by waiting, they think the same thing will happen again. You will see: nothing will happen if we don't make it happen."

Duchâtelet was cold, tired, hungry, and silent. Galois noticed with astonishment that he was not interrupted by his comrade.

"I told them, I tried to convince them that battles are never won by waiting for an opportunity which may never come. We ought to carry the cannon out to the people and incite them to fight. Our strategy ought to be active and not passive. We must provoke the people and not wait for them with folded arms."

Duchâtelet still did not answer. Galois interrupted the annoying silence:

"What do you think? Am I right?"

"We have so many good brains here. Why should I think? Let Cavaignac, Bastide, Raspail think. I shall do my duty. Duchâtelet is for doing dirty work. I am glad that they're doing the thinking for me and taking the responsibility. What bug has bitten you? You are an artilleryman and you must listen to your officers. Then listen to them. Instead you talked to them all day about changing our plans. Why should they listen to you? Who are you? A young man who appeared in a uniform two weeks ago. What right have we to teach them? We are just two young men and not everyone is sure that we are all right. You know what I mean?"

"Damn it, I do know what you mean. Republicans or not, they all think that wisdom comes with age and experience. Oh, Duchâtelet! The world makes me sick. No one wants to listen to me. I am always lonely."

"Now you are ready to cry, aren't you? And you are wrong, absolutely wrong. When I set eyes on you, I knew that you had brains and I liked you for it, though you sometimes make me sick too. But do you think that everyone must like young men with brains? You think that if someone is a Republican, then he ought to be a wonderful fellow and he has no right to be jealous. He is often just as bad as any other fellow. He only happens to walk on the right side of the fence. Look at Pécheux d'Herbinville. He has brains. But I saw how he looked at you. He doesn't like you. He would like to have the monopoly on brains among the young men. He likes me because I am not dangerous, because he is a better speaker than I am, and because he will always be more important than I am. But with you it is different. You may overtake him. See? There's one thing you don't know and that's human nature."

Galois interrupted:

"Human nature! From what I know and from what I've experienced, I detest it with all my heart. I have seen how it works at Louis-le-Grand, at the Normal School, the Polytechnical School, the Academy, and even among the Republicans. I love people collectively, but with a very few exceptions I hate, I detest, I loathe, I despise most of them individually."

"No, you don't," murmured Duchâtelet.

"Oh! You don't know how I suffer, my friend. I hate myself for the hatred that grows in my heart. It was put into my heart by teachers, examiners of the Polytechnical School, Academicians, kings. And here it grows and grows. I could only tear it out from my heart together with my heart and with my life."

Duchâtelet looked at Evariste's tense face, afraid that he might burst into tears. He said gently, "I understand, Evariste. Your real friends know you and like you as you are."

They both remained silent. When the hour of their sentry duty was over, they went to the guardroom. It was warm inside, and there was the smell peculiar to guardrooms all over the world: a compound of sweat, leather, brandy, wine, and dirt.

Galois sat in the corner, took a piece of paper and wrote. When he finished, he went quickly into the center of the room and sprang onto a table where a few artillerymen were playing cards. He disturbed their game and knocked some of the cards on the floor. One of the players shouted, "You bastard, get away! Don't you see we're busy?"

Galois shouted as loudly as he could, "Artillerymen! I want to read you a proclamation! 'To arms! To arms! . . .' "

"Shut up, we've had enough of you for today. We've had plenty of you."

Someone went to Galois, grabbed the paper out of his hand and tore it to pieces. Galois sprang from the table and threw the attacking artilleryman to the floor with the momentum of his jump. The artilleryman tried to turn; they struggled, locked together on the floor.

Suddenly the door opened and a violent voice cried out, "We are surrounded by the National Guard and troops of the line."

Bastide commanded, "Out with your muskets, all of you." Then turning toward the two struggling men, "Stop it. Enough for today."

They both stood up quickly as though nothing had happened between them, took their muskets, and went into the yard.

The Louvre was indeed tightly surrounded. The core which could have infected Paris with revolution was now well isolated. The gates of the Louvre were closed. Only by bloodshed could the artillerymen leave the Louvre, and only by bloodshed could the National Guard enter the Louvre.

Then a loud cannon shot was heard.

The cry "To arms!" resounded over Paris. The Republicans mingled with the people, directed the crowd toward the Louvre. But when they arrived they met a double ring of National Guardsmen and soldiers. One, the inside ring, faced the Louvre. The other, the outer ring, faced the people trying to stream into the Louvre from outside. The people did not attack the National Guard; and the National Guard, in the outer ring, did not attack the people. In the inner ring, the National Guard did not attack the artillerymen in the Louvre, and the artillerymen did not attack the National Guard. Only accusations and shouts were hurled.

"Death to the Ministers!"

"You defend criminals!"

"You fought with us in July and now you are against us!"

"You are rebels!"

"You are bastard Republicans!"

The artillerymen expected to be attacked any moment. They were ready, and those who slept did so with muskets in their hands.

Day broke. Everyone was tired. It was gray and cold; damp snow fell. Thoughts of food, bed, Christmas, sleep, became stronger than the thought of revolution. The two rings of the National Guard became less rigid. Winesellers, butchers, bakers, infiltrated through these rings and sold their products to the artillerymen. They were handed to the soldiers between the iron gratings and money was handed back through the iron gratings. The tragedy became funny, more people filtered through the two rings, and lively conversation was heard between the artillerymen in the Louvre and their friends, sweethearts, and wives outside.

On December 22, the atmosphere in Paris was still tense. By now, everyone knew from the newspaper or from his neighbor that the ministers were condemned to life imprisonment and not to death. Drums beat on every corner. Any small event might disturb the equilibrium and start the Revolution rolling. Proclamations calling for order appeared on the street. They were signed by Lafayette, and their effect was small.

On the morning of this day something happened to tip the scale. The King and the courtiers remembered in time the role that the students had played during the July days. They remembered in time the halo of glory with which the uniform of the Polytechnical School shone in the eyes of the people. Now the directors of the schools were asked to appeal to the students:

"Go out onto the streets! All of you! Go out and urge moderation. Assure the people, as we assure you now, that its liberties will be preserved and watched. Do your glorious duty; prevent bloodshed—in the interest of humanity, in the interest of the people, and in the interest of all France."

The appeal worked. The students of the Polytechnical School and the students of other schools went out, this time with the blessing of their masters. They talked, persuaded, and repeated to the people the assurances which had been given them: that liberty would be preserved. The people, cold, tired, opposed by the National Guard, not supported by the students, isolated from the Republican leaders enclosed within the Louvre, were losing their lust for fight, and dispersed. Then the National Guard dispersed. Then the gates of the Louvre were opened and the artillerymen dispersed too.

Lafayette had done his duty by the King. The National Guard, commanded by the old general, had defended the King and the regime. Order reigned in Paris; no blood was shed.

According to the rules of the game, the old general ought now to demand a reward for his deeds. But Lafayette, with his powers further increased, might become dangerous to Louis Philippe whom he had served so loyally.

The Chamber of Deputies could not afford to fight Lafayette openly. Instead, it performed a magic trick by abolishing the title of the Supreme Commandant of the National Guard. It was not Lafayette who was dismissed. He was not pushed out of the chair in which he formerly sat. No! Only the chair was removed.

The grand old man, the hero of two worlds, was tricked. His vanity was flattered until he found himself serving a policy that was never his. Then, when he had done his duty, when he no longer seemed indispensable, his service was turned off and the National Guard, sword of the bourgeoisie, torn from his hands. On the last day of the year, 1830, the King issued an order disbanding the artillery corps of the National Guard. Thus was the sword torn from the hands of the Republicans.

2 : January 13, 1831

In early January, the following announcement appeared in the *Gazette des Écoles:*

> Evariste Galois, former student of the Normal School, will give a course in algebra designed for young students who, knowing how incomplete is the study of algebra in colleges, wish to examine this branch of mathematics more thoroughly. The course is composed of theories, some of which are new and none of which has been published or lectured upon publicly. Here we shall mention only a new theory of imaginaries, the theory of equations solvable by radicals, the theory of numbers and elliptic functions treated by pure algebra.

The classes will take place every Thursday at 1:15 P.M., at Caillot's bookshop, rue de Sorbonne 5. Beginning of the course: Thursday, January 13.

About forty listeners came to Galois' first lecture. Some were former students of the Normal School who wished to see again the queer young man whom the school had expelled. Some were Galois' Republican friends who came to swell the audience. There was Chevalier who had given Galois the idea for this course, in the hope that some mathematicians would come, that they would understand Galois' work and spread his fame. But no mathematicians came. There were only a few students hoping for an interesting lecture in college algebra. Finally, two police spies completed the strange mixture.

The room adjoining M. Caillot's bookshop was stuffy and smelled strongly of old books; meager light falling through small, high windows illuminated the dust and the old wooden benches. The transitions from light to shadows were sharp, disappearing and reappearing to the rhythm of the passing clouds. Here the greatest mathematician then living in France chose to lecture on his theories to all who wished to listen.

When Evariste entered this room, he was astonished and pleased with the unexpectedly big audience. But when he looked for a new and eager face, he saw none. He caught only Chevalier's encouraging glance and answered it with a weak smile. Then he began his well-prepared lecture:

"Of all human knowledge, we know that mathematics is the most abstract, the most logical, the only one which does not appeal to the world of our sense impressions. Often one concludes that mathematics is, on the whole, the most methodic, the most coordinated branch of science. But this is an error. Take any book on algebra, whether a textbook or an original work, and you will see in it a confused mass of propositions, whose rigor contrasts strangely with the disorder of the whole structure. It would seem that the ideas are so precious to the author that he abhors the pain of connecting them with each other, while at the same time his mind is so exhausted by the concepts which form the foundations of his work that he cannot produce one single thought that would coordinate this ensemble.

"Sometimes you seem to encounter a method, a connection, a coordination. But all this is wrong and artificial. You will find divisions of material for which there is no reason, arbitrary conections, conventional arrangements. These faults, still more glaring than the absence of all method, you will find chiefly in books written by men who do not know what they are writing about.

"All this must seem especially astonishing to people for whom the word 'mathematics' is synonymous with 'rigor.'

"One will be more astonished if one reflects that these mathematical works are destined for the search after truth rather than knowledge.

"Indeed one understands that a mind which could perceive in one grasp all mathematical truth, not only what is known to us but all the truth possible, such a mind could deduce all this truth rigorously and mechanically from a few principles combined with a uniform method. Such a mind would have none of the difficulties encountered by the scientist in his research. But a scientist must work differently. His task is harder and, therefore, nobler.

"The march of science is not along a straight road. Science develops along a strangely shaped path, and in its progress accident does not play a minor role. The life of science is primitive, crude, and disordered. This is true not only for science as a whole, being the result of the work of many scientists, but also true for the particular research of any single scientist. When creating, he does not deduce; he combines, he compares. He does not arrive at the truth, he hits upon it as though by accident.

"In each epoch we find certain characteristic problems. These are the problems which occupy the best minds. It happens that the same ideas are formulated at the same time by a few scientists as though by a revelation. If we look for a cause of this strange fact, we are led to the work of other, earlier scientists where we find the source of these ideas even if they were explicitly unknown to them.

"Science has not profited much from the coincidence of the same ideas appearing at the same time in different minds. Angry competition and a degrading rivalry have been the principal fruits. One can justly conclude, therefore, that scientists, like other people, are not made for isolation, that they are bound to their epoch and they could strengthen their achievements tenfold by association. Then science would develop more rapidly.

"Many questions of a new character occupy the mathematicians of the present day. We shall devote our attention to some of them.

"I shall here present what is most general, most philosophical in my research which a thousand hindrances have kept me from publishing until now. I shall not complicate my presentation by examples or digressions in which mathematicians often drown general concepts. I shall present them always in good faith, indicating in a straightforward manner the way I obtained them and the obstacles I had to overcome. In this way the listener will be instructed in the same way that I was. If I succeed in doing this, I will have a good conscience for having done well, if not because I enriched science, then at least because I gave an example of good faith which is so rare nowadays."

After this introduction, he became technical. But even the introduction was hardly understood. Most of the listeners were so astonished by this nineteen-year-old boy speaking with the air of a great scientist, so sure of himself, so critical of others, that they wondered whether he were crazy or a genius. When later they understood nothing of what he said, they concluded, conveniently for themselves, that not even the speaker knew what he was talking about.

The next week, only ten listeners came; the third week only four. This was Galois' last lecture.

3 : January 16, 1831

It was upon Chevalier's insistence that Evariste followed Poisson's advice and wrote a new manuscript for the French Academy. He looked once more through the eleven long pages. "What will happen to you now, I wonder." He found this thought funny and smiled. Then he rewrote the title page and introduction from the draft he had before him:

ON THE CONDITIONS FOR SOLVABILITY
OF EQUATIONS BY RADICALS

This paper is a summary of a work which I had the honor to present to the Academy a year ago. This work not having been understood, the propositions which it contained having been undoubtedly questioned, I shall content myself with giving here in synthetic form the general principles, and only one application of my theory. I appeal to the referees to read at least these few pages with attention.

The reader will find here the general *condition* which must be *satisfied by all equations solvable by radicals* and which, conversely, assures their resolvability. An application is made only to equations whose degree is a prime number. Here is the theorem given by our analysis:

For an equation of prime degree, which has no rational divisor, to be solvable by radicals, it is necessary and sufficient that all its roots be rational functions of any two among them.

Other applications of the theory are particular theories in themselves. They necessitate, moreover, the application of number theory, and of a particular algorithm: we shall reserve this for another occasion. In part, they are related to the modular equations of elliptic functions, which, as we shall show, cannot be resolved by radicals.

He then wrote the date, January 16, 1831, signed his name, and the same day a new manuscript by Evariste Galois was sent for the third and last time to the French Academy.

4 : February 14, 1831

The parish priest of St. Germain l'Auxerrois was a very old man. He had accompanied Marie Antoinette to the scaffold, and tears had coursed down his cheeks when her head fell. Now, on February 14, 1831, on the anniversary of the assassination of the Duc de Berry, the old parish priest was to hold a requiem Mass for the peace of the dead duke's soul.

Father Paravey, of St. Germain l'Auxerrois, was a young man. He had blessed the tombs of the people who died during the three glorious days, and he refused to be present during the requiem Mass for the dead duke's peace of soul.

A queue of brilliant equipages lined up before the church of St. Germain d'Auxerrois. The aristocracy came to the requiem Mass to show their devotion to the memory of the dead duke, to his exiled father Charles X, and to the duke's son, the lawful King of France.

As he stood among the crowd that watched the exhibition of richness and sorrow, Galois stared at a young girl who stepped from a splendid equipage. She wore a black gown and a black silk cape that outlined her form while pretending to conceal it. Black bows adorned her simple bonnet. Her walk had the air of commanding dignity that forced passers-by to move aside and stare. As she mounted the church steps, her cape fell open at the throat, revealing the whiteness of her skin and the ripeness of her breasts. On such a background, the diamond cross that hung about her neck was not a symbol of religious piety, but a source of sparks that illuminated the rich curves of her neck. Before entering the church, she turned and looked at the people. Her face was angelic, but the half-closed eyes were arrogant and defiant. They wandered quickly from one face to another and then, Galois could swear that this was true, they rested on him for a long time.

He felt how this impertinent angel set his face on fire; how the fire spread downward and made his blood boil. The tension in his body and his mind increased, creating thoughts, scenes, pictures, which in turn increased his tension. He imagined his hands reaching for the cross, pushing the black dress gently aside and then tearing it violently, touching her skin and caressing her breasts.

When the girl vanished inside the church, Evariste felt empty and guilty. He, who understood the problems of algebra better than any other man living, could not understand how a girl who went to the requiem Mass for the Duc de Berry could upset the equilibrium of his mind and body. Did it not mean that Republican ideals were not deeply enough engraved in his heart?

The service in the church began peacefully. But then one of the exquisitely dressed men started a collection for the benefit of the King's soldiers wounded during the July days. Then someone hung up a lithographic portrait of the Duc de Bordeaux, and someone else placed a crown of leaves around this portrait. The religious service drifted into a political demonstration; the first one staged by the aristocracy since the Revolution.

The crowd outside the church increased and its patience decreased. Men came out of the church at short intervals, bringing news of what was happening inside, coloring and exaggerating it with vivid words and broad gestures.

But the crowd that listened now was different from the one which more than a month ago had cried "Death to the ministers!" It was now densely sprinkled with men in black coats and yellow gloves. Once again the bourgeoisie stood united with the poor in whose hearts old hatreds waited for a new release.

A man came out of the church. He climbed on to a horizontal bar of the railings, and spoke:

"Citizens! The aristocrats dare to celebrate a requiem service for the Duc de Berry, for a member of the Bourbon family which we have just driven from power."

"Shame! Shame!"

"Down with the Jesuits!"

"They dare to celebrate this service here in this church near the Louvre which we took, and only fifty yards from where the victims of the Revolution lie buried."

"Shame! Shame!"

"They collect money for the soldiers who killed the people."

"Death to the Carlists!"

"Death to the Jesuits!"

"Shall we allow them to make a mockery of the people's rights that we ourselves have won?"

"No! Down with the Jesuits!"

"Down with the church!"

The crowd attacked.

The church was taken by storm. Some Carlists were beaten up and thrown out of the church; others fled in panic. The altar was pulled down, the pulpit broken, the balustrades and the confessionals thrown on the floor in pieces; the sacred paintings were torn, the rich, golden-flowered hangings trampled by angry feet. All this happened in one moment. The crowd yelled and laughed. They challenged each other to more and more daring acts. Each one wanted to exhibit a courage greater than his neighbor's by committing acts more violent, more vulgar, than those he saw. Priests were cursed; blasphemous cries were shouted. The sacristy was taken; its rich treasures destroyed. One of the rioters came out of the sacristy in the dress of a priest and delivered a mock service to the loud laughter and applause of the mob. But the people halted respectfully before the door of Father Paravey's apartment. They did not forget that he had blessed the victims of July.

When he entered the church, Galois looked around triumphantly. What he saw now was wild devastation, chaos, aimless destruction, sudden outbursts of hatred against furniture and lifeless objects. Triumph changed quickly into shame and humiliation. The Republicans would now be blamed for the sacrilege committed by the crowd, for the excesses in which it indulged, for the devastation it caused. And the royal hands of Louis Philippe would appear clean and innocent. Once more the men with the yellow gloves claimed a common cause with the people, and once more they deceived the people.

He despised himself now more than he despised Louis Philippe and the men who roared wildly through the church. He cursed himself and he cursed the blue impertinent eyes of the girl who had entered the church. He cursed the cross on her chest and the fullness of her breasts covered by the black dress. He knew that he had wanted to enter the church to see her dress torn to shreds, her arrogant eyes frightened and asking for mercy. But he looked for them in vain.

The Prefect of Police reported to the King what had happened at the church St. Germain l'Auxerrois, and the King asked him to stay for dinner. When he learned that the crowd intended next day to attack the Palais-Royal and the Archbishop's palace, he said cryptically to the prefect, "Think only of the Palais-Royal."

And the Prefect of Police understood the King. Next day the Archbishop's palace was reduced to ruins.

5 : March, 1831

The distance between the people of France and the King of the French increased with every day. Soon the government of Louis Philippe would no longer represent even the small bourgeoisie. Soon it would represent only the rich and powerful of France. Soon the people and the bourgeoisie would unite once more in a common victorious cause. And again the people would be deceived and betrayed.

All this was to happen soon: in seventeen years. Very soon, if judged by history and the perspective of the many years past and the many years to come. But it was not very soon if judged by the span of a human life.

But in the year 1831, Louis Philippe believed, as had Charles X, as had Louis XVI, that the end of his reign would come only with death —and that after his death his son, then his grandson would ascend the throne, and that the new line of Bourbons would reign forever over France.

The banker Laffitte left the government and the banker Perier became the King's Prime Minister. Casimir Perier, the strong man of the July monarchy, was tall and looked impressive. His manners seemed calm and noble to those who did not witness his sudden fits of frenzied choler. He hated the aristocracy. He did not hate the people; he despised them. They were a horde of barbarians ready for pillage and happy when swimming in blood. His pride was without bounds. From the heights of his pride, which he did not try to conceal, he looked down upon his cabinet ministers, whom he humiliated with outbursts of anger and irony. There was no generosity, no devotion in his heart, and no loftiness in his mind. He could trample brutally upon his enemies if this was needed to save the power of the bourgeoisie or of the King whom he despised and loyally served.

But to make the King and the regime strong, the power of the Republicans must be crushed. It was strong and becoming still stronger. The Republicans hated the King and, what was still worse, they laughed at him unmercifully. They called him a parrot, a pear (after the shape of his head), they created insurrections and rebellions, they incensed the people, they demanded suffrage for every citizen, they tried to push France into wars for the defense of Poles, Belgians, and Italians; they threatened terror and pillage. France would have no authority, no dignity, no moral strength, no order and prosperity, until the Republicans were crushed. Thus, the great strength of Casimir Perier was bent upon its first and most important task: to break the power of the Republicans.

Two and a half months had passed since, for the third time, Evariste had sent his paper to the French Academy. When he inquired about the fate of his manuscript, he was told that it was being studied by the referees, Messieurs Lacroix and Poisson.

Lacroix was then old, and now his name is of no great importance in the history of mathematics. Poisson was a little fellow who always behaved with great dignity and would never throw a mathematical manuscript into the wastepaper basket. But he was essentially an applied mathematician not much interested in the problems of algebra. There were no great and well-known mathematicians in France at that time. Cauchy had followed the Bourbons into exile, though his presence in Paris would have been of little help to Evariste.

Smiling proudly and bitterly, Evariste whispered to himself, "There is only one great mathematician living in France. And I am the only one who knows who he is."

On March 31, 1831, Galois wrote to the French Academy:

I dare hope that Messieurs Lacroix and Poisson will not consider it in bad taste that I recall to their memory a paper on the theory of equations which they were charged to referee three months ago.

The results contained in this paper are a part of those which I submitted a year ago to the competition for the prize in mathematics and in which I gave, in all cases, the rules for recognizing whether an equation is or is not solvable by radicals. Since this problem has appeared until now, if not impossible then at least very difficult to mathematicians, the referees judged *a priori* that I could not have solved this problem because I am named Galois and further, because I was a student, and I was told that my manuscript had been lost.

This lesson ought to suffice me. Nevertheless, at the advice of an honorable member of the Academy, I rewrote a part of my paper and presented it to you.

Please, M. Président, relieve my concern by asking Messieurs Lacroix and Poisson whether they have lost my manuscript or whether they intend to give an account of it to the Academy.

Accept, M. Président, the homage of your respectful servant,

Evariste Galois

6 : April 15, 1831

Nineteen members of the dissolved National Artillery Guard were arrested and accused of conspiring against Louis Philippe in the month of December, 1830, during the trial of Charles X's ministers. According to the *procureur,* they had intended to deliver cannon to the people, provoke a revolution, and overthrow the monarchy.

Cavaignac, Guinard, Pécheux d'Herbinville, and sixteen others were accused. The choice seemed haphazard, since some of the most active artillerymen remained unmolested. The government wanted first to establish a principle to prove that conspiracy was punishable, before prosecuting other Republican leaders.

Galois went to the Palais de Justice. It was surrounded by municipal guards, and the courts of the Palace under the arches were filled with cavalry. Evariste wormed his way through uniformed lines, through a crowd of workers and students; he had to show his ticket over and over again before he entered the densely packed courtroom, full of pretty and fashionably dressed women. Not one of them noticed Evariste. They all looked with shining eyes at the nineteen heroes.

Galois listened with feelings of sympathy and unity with the accused. Yes, these feelings were in him. Their cause, the cause of the nineteen, was his cause too. But at the same time he was disturbed by an overtone that leveled his enthusiasm, creating a feeling of blame and guilt.

He found himself thinking not only about the accused, their fate, and the outcome of their trial, but also about himself. Everyone had thrown obstacles in his way. He had been persecuted at Louis-le-Grand and at the Normal School; he had been persecuted by the French Academy; he had been persecuted by the Polytechnical School; and he had expected to be persecuted as a Republican. But the last persecution, the only one that brought glory and fame—only that one was denied him. Why was he not accused with the others? Had he not been in the Louvre on December 21? Was he not ready to join the people and overthrow the regime of that man? No! It was not true that the government did not persecute him. It persecuted him most cruelly; it persecuted him by ignoring his existence!

He tried to suppress these emotions and to focus his attention on the trial. He looked at the court. He heard the presiding judge, M. Hordouin, as he turned to Pécheux d'Herbinville.

"You are accused of having arms in your possession and of distributing them. Do you admit the fact?"

The judge pointed to the table on which lay the cartridges seized at the home of the accused. They were wrapped in tissue paper and ornamented with rose-colored favors.

Galois waited tensely for Pécheux's answer, but at the same time a bitter thought crept in:

"What would I answer if this question and the eyes of all these women were directed at me?"

Pécheux looked at the judges and then at the jury. His eyes were as cool and icy as on the July day when Galois had seen him for the first time, when he spoke in the square before the City Hall. With an even and penetrating voice, with a slightly curved lower lip, he answered, "I not only admit this fact, your honor, but I am proud of it. Yes, I had arms and plenty of them too! And I shall be glad to tell you how I got them. In July I took three posts, one after the other, at the head of a handful of men in the midst of the firing. I took arms from the soldiers I defeated. I fought for the people and the soldiers were firing at the people. Am I then guilty if I took away the arms which were intended to wound and kill the citizens?"

Applause greeted these words. Galois applauded too, and whispered to himself, "My time will come. You will not be able to deny me this tribune! You will not be able to ignore me!"

Then came the moment for which the audience had waited. Cavaignac was asked whether he admitted his guilt.

The orator and hero of the people looked at the jury, then at the public, and turning with a magnificent gesture toward the prosecutor, said, "You accuse me of being a Republican. I hold this accusation to be both an honor and a paternal heritage. My father proclaimed the Republic from the heart of the National Convention in the face of all Europe. He defended the Republic; he died in exile after twelve years of banishment. My father suffered for the cause of the Republic which so many others have betrayed. This suffering was the last homage his old age could offer to the country which he had so bravely defended in his youth. And his cause colors all my feelings as his son. The principles in which he believed and for which he fought have now become my heritage. Study, life, experience have only strengthened my convictions. I pronounce it without affectation, without fear, from my heart: I am a Republican."

Galois felt that the disturbing overtones were disappearing, that they were melted by the warmth of Cavaignac's eyes and words. Now he felt his unity with the audience, the unity of emotion which brought tears of love and admiration to many eyes.

"You accuse us of conspiracy. It is an idle accusation. Conspiracies count for little. Revolutions are neither made nor won by conspiracies. The anger, the determination of the people create revolutions. We Republicans believe in the people. To conspire would mean to lose patience, to lose faith in the people. Of this the Republicans are not guilty. We did not conspire. No! It is not we who conspired."

Raising his voice, he pointed toward the prosecutor.

"It is the monarchy that conspired against the people. It conspired by design when it issued the ordinances, it conspired in the past, and it will conspire in the future.

"We Republicans do not need to be overhasty. Our time must come, and our time will come. The world is tormented by new and powerful wants. The people of the world are on the march! Those who pretend to govern are cutting the branch on which they sit; they are destroying the source of their own power. Their deeds—and not conspiracies—create revolutions. Soon even a god would find it more difficult to govern our country than to change it, reconstruct, rebuild it. The bloody deeds of '93 have been cast in the teeth of the Republicans thousands of times. But men with minds, men with hearts, men who love France know that it was the convention that defended the sacred soil of our country; it was the convention that extended France to her natural frontiers, and it was the convention from which all great political ideas have come.

"The Revolution! You attack the Revolution! What folly! The Revolution includes the whole nation, all the people, with the single exception of those who exploit the nation and fatten on the people. The Revolution? It is our country fulfilling her sacred duty of freeing the people who were entrusted to her by Providence; it is the whole of France doing her duty to the world. As for ourselves, we believe in our hearts that we have done our duty to France, and every time she needs us, no matter what she, our revered mother, asks of us, we her faithful sons will obey her."

There was no mere enthusiasm now. The audience became hysterical; men threw their hats, people embraced each other, stood up on the benches, too moved to shout and applaud, too moved to be ashamed of their tears.

Forty-six questions were put to the jury. On the last day of the trial, at a quarter to twelve noon, the jury retired to the consulting room. At half past three, the signal was given that the jurymen had finished their session. The hall was overcrowded with spectators. Thousands waited outside for the verdict.

The foreman said, "Upon my honor and my conscience before God and before men, the answer of the jury to the first question is no, the accused are not guilty. The answer to the second question is no, the accused are not guilty. . . ."

He repeated the words "not guilty" forty-six times. With each repetition the indicator of joy and excitement rose by one degree until it passed the safety point under which silence could still prevail. With the forty-sixth "not guilty" the silence exploded into shrieks, clapping hands, waving hats. One could see the presiding judge moving his lips but his words could not be heard. Everyone knew that he was setting the accused at liberty. Some of the audience jumped over the benches to shake hands with the nineteen and to embrace them. Others rushed out quickly to tell their friends the good news. Joy and enthusiasm spread from the courtroom to the people on the streets. Here the excitement became so great that the lives of the accused were more in danger now from the love of the people than they had been from the hatred of the regime. The crowds were ready to tear the accused to pieces with their embraces. Most of them succeeded in escaping the ovation when they left the palace unnoticed through a side door.

Pécheux d'Herbinville and four of his friends entered a carriage and told the driver to rush ahead as fast as he could. But they were recognized, the carriage was stopped, its doors opened. The five men were carried by the multitude. They bowed, they waved their handkerchiefs, and the air resounded with cheers and shrieks.

Galois saw this scene. He felt a thin stream of bitterness mixed with his joy. The plant of jealousy, though weak and small, took deep root in his heart. It could not be torn out by brutal force or noble resolutions.

7 : Monday, May 9, 1831

On this day, at five in the afternoon, two hundred subscribers to a banquet in honor of the acquitted nineteen gathered in the long room of the restaurant Vendanges de Bourgogne.

There were assembled those who more than anyone else in Paris hated That Man. If these two hundred had been burned to death or poisoned, the Republican movement would have lost its leaders and its heroes.

The chicken was good, the dessert was tasty, and a bottle of wine stood before each guest. The time for speeches arrived. M. Hubert, the toastmaster, rose. He said that Marrast would be the official speaker and that he would propose a toast to the nineteen. Then Marrast rose. "The marquis of revolution" had refined features and abundant locks. He spoke smoothly and with irony about the regime that by the trial of the nineteen had attempted to show strength and determination and had instead shown its stupidity and weakness. He then raised his glass.

"Citizens! To the nineteen Republicans, who in the most noble fashion, by their words and deeds, uphold the honor of France."

"Long live the nineteen!"

"Hurrah for the nineteen!"

"Long live the Republic!"

Cavaignac answered in the name of the nineteen:

"It was but yesterday that I looked over the *Moniteur,* over the records of those famous days, great labors, gigantic wars, over the whole vast enterprise of the French people for the achievement of its rights. I followed this shining path with which the genius of liberty has marked the last forty years, and the events which shook the earth from pole to pole."

He talked about France, the birthplace of liberty, and about her present-day fight, then he said, "Let us remember, friends and citizens, that at this hour we are not alone. It is not only the cause of France that we represent, that we must and will defend. The cause of all free men is our cause. The cause of the people in Poland fighting so gallantly against the Czar's brutal might is our cause too. Did we help them in the hour of their grave need? Did we have anything more than tears for our old brothers in arms? There is a new proverb in Poland 'God is high and France is far.' Yes! The present France is far from all those who fight for their freedom. It is far from Poland, it is far from Belgium, far from Italy, and from all the suppressed nations all over the world and far, perhaps farthest, from its own people.

"The future of France, the future of all the freedom-loving world belongs to the Republicans."

He raised his glass again.

"To the future of France, may she be strong, glorious, and free, and may she bring freedom to all the oppressed."

The glasses were raised solemnly, and only slowly did the background of chatter and talk return to the hall.

As the bottles emptied, the speeches became shorter and less solemn. Now the toasts consisted of brief slogans, thrown into the air, then caught with "long live" or rejected with "down," while the glass of wine was emptied.

"To the Revolution of '89!"

"No, not '89. To the year '93."

"To Robespierre!"

"Long live the convention!"

"To the mountain! Long live the memory of the men of the mountain!"

M. Hubert looked uncomfortable. These toasts had not been planned and they must not be allowed to go on.

He raised his glass and said, "To courageous Citizen Raspail who declined the cross of the Legion of Honor."

"Long live Raspail!"

Galois said to Biliard, a pharmacy student who sat opposite him, "M. Hubert did not like the toast to Robespierre."

"No, and he was not the only one. You ought to have seen M. Dumas' face when '93 was mentioned. They are respectable people, not like us. I wonder whether a toast to Louis Philippe would make them more angry?"

Galois answered excitedly in a slightly drunken voice, "My dear Biliard, you are right, you are absolutely right. We must have a toast to Louis Philippe."

"You are drunk."

"No, I'm not. I shall propose a toast to Louis Philippe."

"If you're not drunk then you're crazy."

"No, I am not drunk, I am not crazy, and I will drink to Louis Philippe."

"They will be on your neck if you do. I shall join them, so help me God."

"No, no one will dare wring my precious neck, and I shall drink to Louis Philippe, so help me God."

A small but effective chorus arose.

"Dumas, Dumas, we want a toast from Dumas!"

Dumas rose. He had the shining black skin of a Negro and blue eyes. His striking red *gilet* was stained with wine, and he spoke with exaggerated gestures:

"To art! Inasmuch as the pen and the paintbrush contribute as effectively as the rifle and the sword to the social regeneration to which we have dedicated our lives and for whose cause we are ready to die."

"Long live art!"

"Long live Dumas!"

"To the Revolution of 1830!"

Raspail rose. The audience looked slightly more sober.

"To the sun of 1831. May it be as warm as that of 1830 and not dazzle us as that one did."

(Many long cheers.)

"May a new revolution come soon!"

"Soon, soon!"

Suddenly:

"To Louis Philippe!"

Minds sobered; hisses began. Everyone stood up and looked in the direction from which the voice came. Had they exposed a spy, whose tongue was loosened by wine? They clenched their fists in readiness to push these words into the mouth that dared to utter them. Colliding and pushing, they all rushed in the same direction. A dense circle now surrounded the source of the treacherous toast.

Then for a second time: "To Louis Philippe!"

They saw Galois. In his left hand he held a glass of wine level with his heart. In his right hand he held a dagger above the glass, its tip pointing toward the surface of the wine. His two fists were firmly clenched, one around the glass, the other around the dagger. He stood like a statue that had come to life only long enough to pronounce the death sentence twice on the King of the French.

The crowd changed. It ceased to be a crowd. Awhile ago, it was united by common anger toward the man who dared to propose a toast to Louis Philippe. But now the crowd of two hundred split into two hundred individuals.

An actor of the Théâtre Français whispered to his friend, Alexandre Dumas, "Let's go away. This is becoming too dangerous."

Dumas disapproved too:

"This is going too far, much too far. Unbalanced young fellow. One must not threaten the King's life."

They left the room in a hurry.

Pécheux d'Herbinville looked at Galois as though the whole affair were no concern of his and said, hardly opening his lips, "You are a fool."

Raspail smiled at Galois and left the ring of Republicans around him. Many of the guests went out of the room quickly, but more than half of them remained. And those who remained rejoiced loudly. They were happy that they had found a clear expression for their hatred, suppressed and, until now, formulated only in oblique slogans and indirect threats. Here was a gesture as sharp and pointed as the blade of a dagger and as strong as the fist that clenched it.

Some of the Republicans now took from the table knives stained with pieces of chicken and raised their glasses, full of wine, half full, or empty, and, imitating Galois' gesture, cried in a chorus, "To Louis Philippe!"

Others, standing in the outer ring, raised only their fists to different levels as though they held the glass and the dagger, and they shouted, "To Louis Philippe!"

After repeating their cry many times with the same gestures, they sought for a new slogan. Someone provided it:

"To the Place Vendôme!"

They responded, "To the Place Vendôme!"

Over one hundred Republicans marched from the restaurant Vendanges de Bourgogne to the Place Vendôme. They pushed Galois into the first row. When they arrived at the Place, they began anew to shout their old slogan with its threatening gesture. Crowds gathered and looked puzzled at the magic sign of the two raised fists. When its meaning was explained to them, they liked it and joined in its repetition. The Republicans from the banquet, the people gathered at the Place Vendôme, formed one fraternal crowd, drunk with wine and the anticipation of victory. They sang the Marseillaise, then they danced around the Vendôme column and repeated with two raised fists, "To Louis Philippe!"

No one interfered with the crowd. They were as gay and as happy as though their magic gesture had annihilated all the tyranny in the world.

8 : May 10, 1831

The police knew everything: they knew what speeches were made, what toasts were drunk; they knew that the King's life was threatened, and they knew the name of the Republican who did it. They knew who left the banquet in protest and who remained. They knew everything.

The magistrate issued a warrant for Galois' arrest. Early next morning the anticipated visitors came. A police commissioner and a gendarme searched his room and took Evariste to the Prefecture of Police at the Place de Dauphine. The three of them entered a small room of the large drab building. The officer of the peace yawned and, without interrupting his teeth-cleaning, took the warrant from the Commissioner of Police and handed him a receipt. This gesture between two bored men set in motion the powerful machinery of justice upon Evariste Galois.

The Commissioner departed, and the gendarme led Galois through the corridor to a long room full of guards in dark-green uniforms and black caps, with clerks sitting at the tables and prisoners standing before them. Some of them were old, some young, some in chains, all of them wretched, badly dressed, and dirty. Behind a wicket at the end of the room, a man counted money and wrote numbers on a piece of paper. The gendarme pushed Galois gently toward this window. Now he would experience all the details of a procedure of which he had so often heard from his Republican friends. How they liked to compare notes, dwell upon the similarities and differences, give advice to those green Republicans who had never smelled the *gogueneau!* Yes, they were right, the wicket looked like a theater ticket window.

The man behind it asked, "*Pistole* or Saint-Martin?"

Yes, it was exactly as they said. One could pay for a private cell—the *pistole*—or one could join others in Saint-Martin, the horror of this place.

"*Pistole.*"

He paid and took the receipt. Then the gendarme handed Galois over to one of the men in green uniforms. Together they went toward a table and the guard emptied Galois' pockets.

The clerk took a piece of paper and without looking up said, "Name?"

"Evariste Galois."

He wrote the name with two l's, and Evariste did not bother to correct him.

"Age?"

"Twenty."

"Profession?"

Evariste thought for a while. Then he answered, "Tutor."

"Place of birth?"

"Bourg-la-Reine."

"Present address?"

"Rue des Bernardins No. 16."

"Height."

The guard measured Galois, checked his result, and pronounced, "One hundred and sixty-seven centimeters."

The clerk wrote down the number. Then he murmured, "Hair."

He looked up and wrote, "Brown."

"Eyebrows." He looked up and wrote, "Id. Forehead, square. Eyes, brown. Nose, large. Mouth, small. Chin, round. Face—" A spark of interest appeared in the clerk's eyes. He looked puzzled, but wrote with determination, "Oval."

After the formalities at the table were finished, the guard took Galois by the arm, led him through corridors and up and down stairs, opened a door and said, "This is your *pistole*."

Evariste entered the cell. It took a long time before the keys were properly turned and, after many crickcracks, he heard the dying sound of receding steps.

He looked at the small window near the ceiling. Through it he saw a few square feet of the finest blue cut by black bars. A beam of light, sharply visible through the dust, entered the small window and on its journey illuminated the furniture at the opposite side. No, this was not only a piece of furniture; it was a legend. He had heard dissertations on this object, how it tortured the prisoner in the long hours of the day, and how it comforted him in the short moments of need.

The *gogueneau* was of metal and covered one square foot of the floor, was as high as a chair, with a rough wooden cover. On this hot day its stink penetrated the nose, the mouth, the lungs, even if one tried not to breathe. The prisoners swore that the *gogueneau,* though emptied every morning, had not been washed since the day of its creation, which must have coincided with the day of the earth's creation.

Time flows only if one puts thoughts and actions into its frame. Otherwise it refuses to move and stands still. Galois began to measure his *pistole.* He did it very slowly, methodically, and carefully. Had he not learned at Louis-le-Grand and then at the Normal School that every experiment must be performed three times and the average taken? The result of the averaging was eight feet by six feet, or, concluded Galois, forty-eight square feet. One square foot for the *gogueneau* and forty-seven more. Then he began a careful study of the forty-seven.

He looked at the bed. It was a heavy mass of wood; on it was a straw mattress, a dirty cushion, two coarse sheets. He touched them and concluded that they were coarser than the straw in the mattress. Then he touched the blanket and concluded that it was coarser than the sheets.

After completing his examination of the bed, he looked at the rest of the furniture. There was not much to investigate: one chair, a table carved by the work of idle hands and gray with the dust of years. Then the walls! They were covered with signatures and initials, some made by pencil, some by nails. They were accompanied by indecent pictures of men and women, always nude and with certain parts of their bodies made much more prominent than others. Some of the figures were busy making love, others sitting on the *gogueneau.* The signatures and drawings were accompanied by dates and slogans, some of them obscene, some of them revolutionary.

Printed matter hung on the gray wall near the door. Galois studied it very slowly. It was an announcement signed by the director of the establishment. There the price per day of a *pistole* was stated. Then a detailed list of all the furniture. Galois read the items one after another and compared the list with the world of his sense impressions.

"A table. There is a table. A pitcher. Let me see. Yes, quite right. There is a tin pitcher and it stands on the table. Time has eaten it, hands have dented it in a hundred places. The pitcher is a sorry thing but it cannot be denied that the pitcher exists. A chair. Yes. Bed. Yes. Ah, here we have the seat of torture. Why mention it at the end? Why after the pitcher and not before the pitcher?"

The list of furniture covered only a few lines. Below was the legend. It stated clearly, leaving no shadow of doubt, that the *pistole* was to be paid for daily, that the lodger was responsible for the well-being of all the furniture, and finally that the lodger who did not pay for his *pistole* would be transferred immediately to Saint-Martin, the horror famous in the history of prisons.

Galois read the list and legend once. Then he began to read it again. After that he knew everything by heart and even had an exact picture of all the dots with which the flies had ornamented the director's announcement. There was no use reading it for the third time. He might just as well sit in the chair and repeat it by heart. He took off his coat, then his shoes. With some relief he found the smell more bearable. He intended to write his name on the wall, but started to think about elliptic functions instead.

Mechanically he lay on the wooden bed. He felt an irritation on his left arm. When he scratched, the irritation moved upward. He looked at his bed and saw on its cover a bug, small, dark-reddish, flat, creeping ahead very slowly. He killed it, and his finger became reddish with a messy, squeezed piece of bug flesh. The smell was so sharp that for a moment it seemed to drown even that of the *gogueneau.* He took off his shirt and looked for the bugs on his shirt and body. He found two of them.

"This," Galois thought, "is one way of filling the frame of time with action and making it flow."

Soon his meal came. A small window was opened in the door, and a plate of beans, a pitcher of water, and a wooden spoon were pushed in. A voice announced through this opening that he might send for his own food if he wished to pay for it. The thought of eating in this atmosphere of the mixed smells of the *gogueneau* and dead bugs made him sick. He drank the water and left the beans untouched.

He lay down on the bed. An idea about the connection between algebraic equations and elliptic functions occurred to him. A few minutes later he forgot where he was. Mechanically his hands scratched the irritated spots and dispersed the flies. Now he was far away from his cell; even its smells ceased to disturb him.

9 : May, 1831

Next day, Galois was sent from the Prefecture of Police to the prison Sainte-Pélagie. With eleven others, he stood before the *panier à salade.* This was the prisoners' name for the truck which transferred them from one lodging to another.

The *panier à salade* looked clean and shiny from the outside. A gendarme helped the prisoners, not too gently, to climb to the high level of the truck floor, and then closed the door.

Inside it was dark. A little light came from the grating at one end of the truck. Through it, Galois saw two uniformed backs and some tiny segments of horseflesh. The long parallel walls had four small holes, each ten centimeters in diameter. The benches which ran along these walls formed inclined planes with their lower edges toward the interior of the truck. Like everyone else who rode in the *panier à salade,* Galois wondered why the benches were made in this fantastic way. Neither he nor anyone else could find the answer. Was it, perhaps, to occupy them with the task of keeping their equilibrium so that they could not chatter to each other? Indeed, each of them silently braced his hands against the knees of the prisoner sitting opposite.

Through the tiny holes, Evariste recognized the walls of Louis-le-Grand as the truck passed them. For the first time, the recollection of Louis-le-Grand seemed pleasant; a peaceful world which had passed forever; walls that had sheltered him from the outside world, more cruel and dangerous than anything the masters of Louis-le-Grand could create or imagine.

The noise of the moving truck mingled with the pleasant ringing of the bells on the horses' necks. The bells announced to the citizens of Paris that the enemies of the state were passing and that theirs was the right of way over the carriages of the rich.

The truck arrived at Sainte-Pélagie. It stopped at the rue du Puits-de-l'Ermite, before the entrance to the prison. Both the gendarme and the postilion came down from their seats and opened the iron door. Another mounted gendarme who had followed the truck along its journey grimly watched the debarkation of the three political prisoners destined for this place of detention. The postilion helped them jump down and then stretched out his hand, shamelessly demanding a tip for his labors. The small windows of Sainte-Pélagie which looked out into the wretched rue du Puits-de-l'Ermite were now full of faces pressing between the squares made by iron bars.

"Welcome, welcome to your new home."

"Hurrah for our new patriots!"

When the new prisoners entered the building, its old inhabitants assembled in the yard to greet the newcomers. Galois was received with shouts of joy and with the threatening gesture of two raised fists. The prisoners had read and had heard about the banquet. Now they asked Evariste to repeat again and again all the details, everything that had happened in the restaurant Vendanges de Bourgogne and at the Place Vendôme.

The prison Sainte-Pélagie was divided into three isolated parts. The one with its entrance on rue du Puits-de-l'Ermite was for political prisoners only. Here in the big yard the prisoners could walk freely, talk, discuss politics, and quarrel with each other. Or they could go to the canteen, a dark and dirty room where they could write, buy food, play checkers, get drunk, discuss politics, and quarrel with each other.

They were little watched and could express their hatreds aloud, and boast still louder about the deeds that had brought them here. Often their stories fell upon the sympathetic ears of spies who knew how to pretend friendship and extract confessions.

In the evening, the prisoners went back to their cells, which were closed by the turnkeys and reopened early in the morning. There were some small cells for a few prisoners, and large cells, each of them with about sixty beds. Many of the cells communicated by doors that were shut for the night.

No one had privacy, no one was left alone. Newspapers and visitors brought with them the reflection of the outside world. The newcomers were asked questions without end: How was Paris, how were the Republicans, was there hope for a new revolution? What brought them to Sainte-Pélagie? The prisoners knew all the answers beforehand, but like children with unlimited time and nothing to fill it with, they listened eagerly to the repetition of old tales, spitting when the name of Louis Philippe or Casimir Perier was uttered, nursing hatred, and increasing their hope of revenge.

Besides the political prisoners (and spies pretending to be political prisoners) there were, at Sainte-Pélagie, two hundred and fifty children between ten and twelve years old. These were the abandoned children whom no one claimed and no one loved, rounded up on the streets of Paris like dogs without masters. Why were they put among the political prisoners? True, they joined every revolution and every rebellion. True, their courage had disarmed soldiers and taught grown-up men how to fight. But here, imprisoned with their elders, playing at battles in the courtyard, listening to political discussions, they learned to fight with still more determination and hate with still more violence.

A man and his wife were hired by the prison authorities to take care of these children. The man was kind and he taught them singing, writing, and reading. The man's wife was kind and mended the children's rags. To this kindness they responded with doglike devotion and all the accumulated love which they had had no other chance to bestow.

The children went to their cells early in the evening. Then the prisoners held their daily "service." The tricolor flag was placed in the middle of the courtyard, and the prisoners, their heads bared, surrounded their symbol of free, Republican France. They all sang the Marseillaise. The children pressed their small faces between the bars of the grilled archways that closed their cells and joined their elders in song. While singing the words *Amour sacré de la patrie!* the prisoners knelt and the guards removed their hats.

No one spoke after the Marseillaise ended. Then the silence was broken by the children's song:

> When our elders have departed
> We shall follow their career
> And from them the light imparted
> And their dust will guide us there.
> Less our zeal is to outlive them
> Than to join them in the grave.
> Ours will be a pride most solemn
> To avenge or share their fate.

The oldest among the prisoners went to the flag and kissed the tricolor cloth. Others followed him. On his first evening in prison, Galois, in his turn, kissed the flag with tears in his eyes. He was too moved to utter a word to his comrades when he entered the cell which the turnkeys closed for the night.

A visitor from the outside world came to see Galois. It was M. Dupont, a well-known Republican lawyer, one of the defenders at "the trial of the nineteen." He told Galois that he would defend him and that he was sent by the Society of the Friends of the People.

His eyes showed sympathy and his smile was patronizing when he said, "The affair is not as serious as it looks. The Orléanist papers try to make a regicide of you. If one reads them, one might think that you have already killed the King. They make a great fuss, saying that you are the first Republican who, since July, has threatened the King's life, that you are a dangerous fanatic who may even do it some day." M. Dupont laughed. "They ought to see you. Then they would know that you would not hurt a fly."

Galois' discouraging look was of little avail. M. Dupont proceeded quickly:

"I want to tell you something important, M. Galois. We must try to wipe out any impression that you wished to assassinate the King. Such an impression would be wrong of course, and very dangerous for you. The King's police would try to remove you by legal or illegal means. Even if the jury should acquit you, your troubles might just begin. You don't need to be frightened, because I don't believe in this danger; I am sure that we shall be able to convince everyone that your life is not a threat to the King's life."

Galois asked, "What do you mean by that?"

"You did not see the articles in the press, which reported the whole incident incorrectly. Republicans who sat near you distinctly heard you say 'To Louis Philippe, if he betrays.' Not everyone heard the last three words, because they were drowned by hisses and loud protests. But we have enough witnesses who sat near you and who heard these words."

Galois looked with disgust at his lawyer and said, "I can't remember saying 'if he betrays.' "

M. Dupont smiled.

"Of course you have no doubt that your comrades speak the truth. They say they did hear 'if he betrays.' It would be very unfair to deny it. It would only mean that you were very drunk and that you don't remember what really happened. This is the view the jury will probably take. Otherwise, they would have to believe you and say that your witnesses lie. Of course you would not like that either. It would mean that your witnesses lied under oath. Do you understand the situation?"

"Yes. I understand the situation."

"I knew you would. I have heard how clever and logical you are. Try to remember everything that happened and how it happened. And then you certainly will remember that you said 'if he betrays.'

Your comrades claim that no good Republican would unreservedly threaten the King's life nowadays, and that this alone is sufficient proof that you must have said 'if he betrays' although only a few heard it. Do you understand, M. Galois?"

"I do understand."

"Of course you do. I knew you would."

10 : Wednesday, June 15, 1831

Flanked by two gendarmes, Evariste entered the hall through a small door at the left of the presiding judge. All eyes turned toward Galois. It was the same courtroom where, two months ago, he had witnessed the case against the nineteen. Yet it looked very different. Two months ago, he stared with fascination and jealousy at the judges, the accused, and the jury. The light from the long, high windows had illuminated the scene in which Cavaignac had given such a masterly performance. It was the hall where twelve men gave a just verdict; an island in France where justice reigned.

But today the charm had disappeared. The hall was dim; the light from the long windows did not penetrate the courtroom. With difficulty he recognized his friends and fellow guests at the Vendanges de Bourgogne. The judge's robe was worn out and dirty. The green cloth covering the table was stained and patched. He did not think about justice and he was not afraid of the verdict. The presiding judge seemed friendly and intelligent, and the men on the jury seemed stupid but harmless. Yet he felt the great burden of being the chief actor on a stage where he had to perform for history, perform without rehearsal, without help. Tomorrow the newspapers in Paris would give an account of what he said and how he said it. Tomorrow the whole of France would know whether Evariste Galois was afraid of Louis Philippe!

The clerk of the court read the accusation in a monotonous voice to the conclusion:

"Galois is accused of provoking, by a statement offered in a public place and a public meeting, an attempt against the life and person of the King of the French without the same having been carried into effect."

Galois looked at the presiding judge, at his small gray beard, at the gray mustache, and the gray eyes which seemed clever and human. The judge began the examination in a kind voice in which there was neither impatience nor animosity.

"Prisoner Galois! Were you present at the meeting which was held on the ninth of May this year at the Vendanges de Bourgogne?"

Galois thought again how different the show looked from the audience than from the stage. There it seemed a drama full of pathos, here, a petty business of small questions and small answers. He had been tricked and forced into admitting that what he had said was, "To Louis Philippe, if he betrays." But he must show the judge, the jury, and the whole world that his hatred was boundless and that he had the courage of his convictions. He must make it clear, clear without any shadow of doubt, that the additional three words were not an expression of cowardice or of a desire to be acquitted by the twelve men on the jury.

The presiding judge waited patiently for an answer and repeated his question in the same words:

"Prisoner Galois! Were you present at the meeting which was held on the ninth of May this year at the Vendanges de Bourgogne?"

"Yes, sir."

"How many guests were there?"

"About two hundred."

"How were you invited?"

"The newspapers announced the banquet, and commissioners were entrusted to examine those who wished to attend. I was among them and I was admitted."

"What was the occasion for the banquet?"

"The acquittal of the nineteen and the refusal of M. Raspail to accept the cross of the Legion of Honor."

"Several toasts were proposed. Could you tell us about them?"

Galois looked defiantly at M. Naudin, the presiding judge, and said, "To 1793, to Robespierre, and others which I don't remember."

"By whom was the toast to 1793 proposed?"

The defiant look hardened and there was a trace of irony in the answer:

"I don't remember."

"Wasn't a toast proposed to the sun of July, 1831, to which was added 'May it be as warm as that of 1830 and not dazzle us'?"

"Yes, sir."

"Who proposed this toast?"

"I don't know."

Now the tone of irony in Galois' voice began to be more distinct with every succeeding answer.

"After this, did not some voices cry out 'sooner, sooner'?"

"Yes, sir, *everyone* said that."

While M. Naudin was looking for the next question, Galois said, "Sir, I was at the banquet and if you will allow me to say what happened there, you will save yourself the trouble of questioning."

The presiding judge looked amazed and said in a friendly voice, "We shall listen."

"This is the exact truth concerning the incident to which I owe the honor of appearing before this court. I had a knife which now lies on the table. I carved my chicken with it at the banquet. After dessert I raised this knife and said, 'To Louis Philippe, if he betrays.' These last words were only heard by my immediate neighbors because of the fierce hootings that were raised by the first part of my speech and the notion that I intended to propose a toast to That Man."

Galois stopped abruptly.

The presiding judge asked, "Then in your opinion a toast offered purely and simply to Louis Philippe, King of the French, would have excited the animosity of the whole assembly?"

"Obviously, sir."

"Your intention therefore was to use the dagger on Louis Philippe?"

Everyone waited in tense silence for the answer. It came promptly.

"Yes, sir, if he should betray."

The judge did not seem shocked. His tone seemed to become a little more friendly. The King's prosecutor looked triumphantly at the jury while M. Dupont tried to conceal his anger by an ironical smile.

The presiding judge asked, "Was it the expression of your personal sentiment to brand the King of the French as deserving the dagger stroke, or was your real intention to provoke others to action?"

Galois answered calmly, "Both. I wished to do it myself and to incite others to such an action if Louis Philippe should betray—that is to say, in case he ventured to depart from legal action."

A murmur of amazement went through the audience. This ceased to be courage. It was pure madness. Whatever happened, the poor, mad boy was sealing his fate. The King's police would remember these words.

The judge looked at Galois with sympathy and asked, "Do you suppose the King is likely to act illegally?"

"Everyone with a little brain knows that it will not be long before he is guilty of this crime, if, indeed, he has not already done so."

M. Dupont sat resigned.

Raspail murmured to himself, "I never saw anyone with such a strong instinct for self-destruction."

The presiding judge asked, "Please explain yourself."

"Is it not obvious, sir?"

For the first time, the judge seemed slightly impatient when he replied, "Never mind! Explain it."

"What I said is that the trend of the government's actions leads one to the conclusion that Louis Philippe will some day be treacherous if he has not already been so. Let us view the facts. It follows from them clearly that Louis Philippe is capable of betraying the nation. Indeed! Let us recall his accession to the throne. Did he not prepare this accession for a long time ahead? Did he not assure Charles X repeatedly that he was his most loyal subject? And then . . ."

M. Dupont interrupted:

"Your honor! Please do not continue the questioning. I admit the ground which M. Galois wishes to cover is dangerous to him, but it is still more dangerous to the King. If the examination proceeds along this line, I shall be forced—much as I would dislike it—to add my own explanation. I possess proof leaving no doubt whatsoever that the accession of Louis Philippe was prepared long before. I will have to present these proofs to the jury."

The judge seemed irritated.

"I have the right to direct the examination as I wish, and I can ask the accused all the questions I believe to be suitable."

Then something unexpected happened.

M. Miller, the King's prosecutor, rose and said, "I join the counsel in begging the presiding judge not to continue the examination in this field."

Amazement spread through the jury and through the audience. M. Naudin turned to the twelve men.

"The jury will understand why I discontinue the examination along this line." Pointing to the knife on the table he asked, "Why did you bring this knife to the banquet?"

"By mere chance. After I bought it I carried it every day."

"Did you order it in this shape?"

Galois seemed amused. He smiled and answered, "Yes, sir. Isn't it a precious instrument? With such knives Republicans carve turkeys and chickens."

By now the presiding judge had had enough. He said, "Thank you, that will be all for the moment."

The parade of witnesses began. The first six were waiters who testified to the general character of the banquet.

M. Gustave Drouineau, writer, decorated with the cross of July, was ushered in.

The judge said, "Raise your hand."

M. Drouineau did not raise his hand and announced in a dignified manner, "Sir! I refuse to take an oath. The records of the pro-

ceedings must have shown you, sir, that I do not feel myself bound, nor am I willing to reveal anything of what happened at this private banquet. I do not intend to defy the law, but, I repeat, the privacy of the place frees me, as far as I am concerned, from the obligation of giving evidence."

The presiding judge explained patiently, "Everyone summoned before the court must give evidence of any knowledge he possesses, unless he is found to be in one of the catagories exempt by law from such an obligation."

M. Drouineau put his left hand upon his heart and said, "I solemnly declare that I shall never consent to give evidence on matters that took place in private. There is a law more sacred than those written on perishable paper and that is the law of honor. The gentlemen of the jury will understand me."

This was an important matter for the prosecutor. He knew that M. Drouineau was Galois' neighbor at the banquet, and that he had left the hall indignantly after the toast to Louis Philippe. It was his testimony which would decide whether the words "if he betrays" were added. Therefore, M. Miller insisted:

"M. Drouineau is obliged to give evidence. If not, he is guilty according to articles 355 and 80 of the Criminal Code."

The quotation of the articles did not frighten M. Drouineau, neither did it disturb his perfect calm nor his excellent manners, as he declared, "When summoned previously in front of the examining magistrate, I was condemned to a fine because of my refusal to give evidence. It seems to me that by virtue of the maxim *non bis in idem,* I cannot now receive a second penalty; the law cannot punish me twice for the same offense."

But M. Drouineau was wrong, as he found out when the presiding judge, after deliberation with the other judges, sentenced him to a fine of one hundred francs.

After that, others of Galois' neighbors at the banquet testified. Yes, they had all heard distinctly that Galois had said, " To Louis Philippe, if he betrays."

It was evening when the prosecutor began his speech in a declamatory voice, with broad and drastic gestures:

"Many punishable crimes have been committed by the Republicans. But never before since July has a Republican dared to threaten the life of the lawful King of the French. Never before, until the ninth of May! On this day, Evariste Galois held up his dagger which, as he confessed himself, he wished to stain with the blood of the King. He confessed here, before you, members of the jury, that he wished either to stain his dagger with the King's blood himself, or to incite others to the greatest crime the human mind can conceive. He had the sad courage to utter his threats in a public place.

"But his crime, mad and dangerous as it appears, is still worse in reality. When asked by the investigating judge, the prisoner Galois admitted that what he had said was 'To Louis Philippe.' But today he changed his tune. He claims now that what he said was, 'To Louis Philippe, if he betrays.' Then it is obvious that he lied once, that he either lied before, or that he lies now. When did he lie? To the investigating judge or to the members of the jury? Is it not reasonable to assume that the accused, in spite of his vulgar boasting, is afraid of the wrath of the people which you will express when you find him guilty? Is it not reasonable to assume that it is this fear of your judgment that compelled him to change his confession? How otherwise can we explain that more than a month after the infamous banquet he remembers better than after a week? The only possible conclusion is that the accused lied to you, that he threatened the life of the King with raised dagger, in three words, 'To Louis Philippe.'

"Gentlemen of the jury! We have before us one of the most dangerous men who ever appeared before this court. He is dangerous to the life of the King which he dared to threaten, and he is dangerous to all those who wish to enjoy the peace and liberty we won in July. And he is doubly dangerous because he is well educated and intelligent, a former student of the Normal School, from which he was expelled for his wicked and immoral character.

"Members of the jury! Only by declaring the prisoner guilty, only by convicting him to long imprisonment shall we show that France cares for the safety of her King. The accused not only dared to threaten the King. He dared, here, in this courtroom to insult the King. He dared to say that the King of the French, who swore to obey the laws of France, is likely to betray his oath. This alone should be enough to convince you that there is only one verdict for this man: guilty! If you do not condemn him, if you do not punish the hand which raised the dagger, then you deliver France to anarchy.

"Members of the jury! Do your duty to the King and to France. The laws of France protect the life of the humblest man. But above all, they must protect the man who is King of the French people. It is your privilege and your duty to show the world that law reigns in France and that France protects her King."

The prosecutor gently wiped his perspiring forehead with a handkerchief, sat down and looked around indifferently. The judge turned to Galois.

"Prisoner Galois, do you wish to say anything in your defense?"

"Yes."

"You may."

"I should like to correct a few of the prosecutor's errors—both in logic and in fact. The prosecutor builds an elaborate theory on the fact that I said different things to the investigating judge and to the jury. The investigating judge asked me whether it is true that I said, 'To Louis Philippe.' My answer was 'Yes.' He did not ask whether I said something else too. Why should I volunteer additional information? If you had seen the investigator's happiness when I admitted that! The man was overwhelmed with joy that he had discovered a great revolutionary. Nothing else could have made him so happy. It would have been cruel on my part to spoil his joy by weakening what I had said and volunteering additional information for which I was not asked. I didn't have the heart to do it. Can you blame me that I didn't want to spoil such great happiness?"

Some of the jurymen suppressed laughs. Some of them watched the prosecutor, who bit his lips. When he saw the looks directed toward him, he quickly improvised an unconvincing yawn.

"Let us take the other argument of the King's prosecutor. How, he asks, can a King err, how can we even conceive that a King can break his oath? But, after all, no one is so childish or foolish as to suppose that a King is perfect. Such silly statements can be heard nowadays only in a courtroom and only from the mouth of the prosecutor. Let us examine this naïve argument a little more carefully. Imagine that a year ago I had said that Charles X would betray. Would not this or another prosecutor have demanded my head in the name of the King who was wise, perfect, loyal, incapable of error or betrayal? But now if I said that Charles X erred, this same public prosecutor would have only praise and sympathy for me. Who then can know what will happen in a year? Perhaps this or another prosecutor will praise me for the wisdom of having predicted Louis Philippe's betrayal. The prosecutor has said that I am well educated and intelligent. I am sorry that I cannot return the compliment. How can any one well educated in history uphold the dogma that kings never betray and never err? As to the intelligence of the prosecutor . . ."

The prosecutor, with badly concealed rage, stood up.

"Your honor! I protest!"

The presiding judge turned to Galois. His voice was calm and friendly:

"You must not offend the prosecutor. I shall not allow you to continue along these lines."

"Thank you, sir. I shall not continue along these lines. I shall change my line completely, and I shall try to please the prosecutor as much as I can. I ought to do at least as much for the prosecutor as I did for the investigating judge. The prosecutor has tried to convince you, members of the jury, that I am one of the most dangerous and most fierce Republicans, that my freedom will be a constant threat to the King and to the government. I shall admit that here he is right. I am a Republican and I am proud to be regarded as dangerous to the regime. I shall admit more. For the last few months I have run about the streets of Paris, always armed, always ready to stir up insurrections, always ready to take part in rebellions. It is only by pure accident that you, members of the jury, see me here for the first time. I was in the Louvre on December 21, last year. You, who accuse me, thought when you took power there would be no rebellions. But you were wrong. There were, and they will continue until you lose power."

M. Dupont stood up.

"Your honor, the prisoner is prejudicing his own case."

The presiding judge turned toward Galois.

"M. Galois, I cannot allow you to prejudice your own case."

The sound of Galois' voice drowned out that of the judge.

"I am finishing. You are childish. You have put our heads on the scaffold, but you lack the strength to let the knife fall. It is we, the Republicans, who stand for strength, courage, progress. Corruption will never reach our Republican souls. It is you, men of the Restoration, who are reactionary and corrupt. We could explain our aims in a way that would confound our accusers; never again would they take our silence for acquiescence."

The presiding judge interrupted firmly:

"In your own interest, I forbid you to continue."

Galois turned toward the judge with sudden calm.

"Never mind, sir. I have finished."

Then M. Dupont spoke. He was not in good form. Galois' mad behavior had upset his prepared line of defense. He discussed at length the legal question as to whether or not the restaurant was a public place. "No," argued M. Dupont, to which the prosecutor replied with a "yes" supported by precedents and law. Again M. Dupont replied with a "no" supported by other precedents and other laws.

Then came the judge's summary. He was not an orator; he was even handicapped by a slight lisp. But his gray eyes blinked with sympathy when he said at the end:

"The case is clear, since the accused does not deny that he made the toast with a gesture threatening the life of the King. Witnesses have confirmed that he said 'To Louis Philippe, if he betrays.' In judging the case you ought to dismiss from your mind the words by which the accused has prejudiced his own case. In judging the accused you may, and I believe you ought, to take his youth into account. He is not yet twenty years old. Some of you may have or may have had sons of this age, and those of you will know that the spirit of revolt passes with age if treated not by punishment, but by persuasion and mercy. I believe that you may and perhaps ought to take this into consideration when debating the verdict."

The jury retired.

M. Dupont said to Galois, "You may have a chance here, but you have harmed your case terribly. What you did is not courage, but madness."

Galois did not answer.

M. Dupont added in anger, "I warned you before. You seem to think that the jury alone punishes or frees. I hope you will never find out otherwise."

Still Galois did not answer.

"The jury is returning."

There was a murmur of surprise. It was only ten minutes since the jury had left the courtroom. A verdict had been reached in ten minutes! No one remembered anything like this happening before.

"The accused is not guilty."

When the presiding judge announced that the prisoner was free, Evariste went to the table, picked up his knife, shut it, and dropped it into his pocket. Then he bowed to the judge, turned ninety degrees, bowed to the jury, and went out without saying a word.

11 : 1831

M. Gisquet, Prefect of Police under Casimir Perier, sat in his office. He looked self-consciously and with embarrassment at the frozen statue on the opposite side of his desk, tossing words violently in its direction, hoping to bring a spark of life into the motionless figure.

"You are right, you are absolutely right, M. Lavoyer. And you are the right man to do it. I shall see M. Perier tomorrow and will ask him for money to organize your work. Yes, M. Lavoyer. Money won't be a problem at all. We must greatly increase your division. Come back the day after tomorrow, after I have seen M. Perier, and I'm sure that you will get the money to organize your work. Then we'll discuss the details of organization."

The man sitting opposite might have been dead. His thin face did not move, and his somewhat slant eyes did not blink.

The lack of response excited M. Gisquet, and his voice increased in violence.

"We have no choice. The jury has acquitted a man who admitted that he intends to murder the King. We must remove these dangerous subversive elements without the help of a jury. We must bore from the inside and from the outside. You're right. We have been much too soft for our own good. We have infested the Republicans with spies. Spies and more spies, that's all. We always knew what happened and who did it.

"Then we took our cases to a jury full of Carlists and Republicans. Much as they hate each other they work beautifully together on a jury. They know how to acquit the enemies of the King. Even the judges are against us. We must change all that. These are all childish methods. They play into the hands of Republicans, giving them a tribune from which they shout to all Paris. They have been swelling their rank, sure that they will remain unpunished. Let them wait awhile. They'll see."

In his excitement, M. Gisquet forgot his visitor. He got up, began to march up and down the office, and his voice grew loud; his gestures grew as violent as though he were addressing thousands.

"We must remove their leaders. We must introduce hatred and discord among them. They must kill one another in duels and bloody fights. Then accidental bullets will kill some of them; no one will know from whom or whence these bullets come. We must have women in our service, pretty women, dangerous women. We shall plant among them jealousy, infidelity, mistrust, hatred. And, by God, we shall get our men by ourselves if we cannot get them from the courts. We shall poison their lives until their courage falters.

"Under pressure from the outside, they'll start to rot on the inside. Once the leaders are gone and lose their authority, the mob will calm down. There will be no one to incite them. And then the power will be in our hands. That's what we want. That's what we must do and that is what we shall do."

He was tired by the outburst of his own energy. He sat down and turned to the silent figure.

"Do you agree, M. Lavoyer?"

Mr. Lavoyer's lips hardly moved as he answered, "What you just said, M. Gisquet, is a very sketchy summary of the report I submitted to you a week ago."

M. Gisquet was suddenly deflated. He stammered, "Yes. Of course. I know. You are right. This is what you wrote." Soon, however, he regained his self-confidence. "I know that you're the right man to do it, but I must warn you. Everything must be done in such a way that even a hundred years from now no one will be able to detect anything. Without documents, without papers, without formalities. If the opposition ever finds out, we are lost."

The wax figure answered, "They never will."

"This is exactly what I wanted to hear from you. But, M. Lavoyer, I must warn you quite honestly that if, at any time, something is found out, I shall refuse to take any responsibility. I shall throw it upon you. You will have a completely free hand. I don't want to know any details. I wish to find out, just as any other Frenchman would, accomplished facts, and I wish to wonder, as any other Frenchman would, whether what happened, happened by design or by accident. And like any other Frenchman I must not be allowed to find out, even if I should investigate and reinvestigate for a hundred years. I don't want to know anything about it. Is that clear, M. Lavoyer?"

M. Gisquet laughed and looked at the man opposite him. His smile froze and he said, "I tell you all this because I am an honest and sincere man."

M. Lavoyer spoke with tightly-pressed lips.

"Yes. You tell me all this because you are an honest and sincere man."

The Prefect of Police looked at the cold eyes opposite him. A disturbing thought came to his mind. These eyes could make even him, M. Gisquet, afraid and unsure of himself. He tried to cover his embarrassment under a friendly, personal tone of voice. But he detected in it an overtone of fear. He wondered whether M. Lavoyer heard it too. This thought made the overtone still stronger and more distinct.

"I wonder why you are doing all this. You cannot expect recognition, you cannot expect fame. Yours will be the most thankless, the most dangerous assignment in our country."

M. Gisquet waited, wondering whether M. Lavoyer would offend him by not answering, and then whether he ought to insist or whether he ought to change the subject. With relief he saw the tightly pressed lips begin to move.

"You, sir, like to have power and to be known. I like to have power and to be unknown."

"Yes, I see, I see. But we can leave this subject. I asked you about the dossier on Galois. You have it?"

M. Lavoyer pointed to the thick dossier lying on the desk.

"Have you a short summary of the case?"

M. Lavoyer opened the dossier, took out two sheets of paper covered with neat handwriting and handed them over to M. Gisquet.

"Yes, it is a pleasure to work with you. Everything always in perfect order."

There was no reaction on the stone face.

"What about reading it aloud?"

M. Lavoyer took the two sheets of paper in his hand. His fingers were long, they looked delicate, and at the same time they were strong and prehensile. He read in a quick, monotonous voice:

"Short History: Evariste Galois. Born at Bourg-la-Reine on October 25, 1811. Father, mayor of Bourg-la-Reine, was a Liberal with Republican sympathies. He was never engaged in any political conspiracies. Mother ambitious, strong, and somewhat queer. One sister, twenty-four years old, and a younger brother, seventeen years old, both not interested in politics. Father committed suicide in 1829. His funeral incited the inhabitants of Bourg-la-Reine to a riot in which the parish priest was hit by a stone. Evariste Galois was one of those who incited the inhabitants to this riot.

"Failed twice in the examination for the Polytechnical School and entered the Normal School. Did not take part in the July Revolution but fought with the director, M. Guigniault, for not letting the students out of the building. On December 3, 1830, wrote a letter against M. Guigniault to the *Gazette des Écoles*. Expelled from the Normal School on the fourth of January, 1831.

"In August, 1830, became a member of the Society of the Friends of the People. Tried to incite the members and spectators at a public meeting of this society on September 17, 1830, by a violent speech and by shouting 'Death to the ministers.' Joined the National Artillery Guard and spent the nights of December 21 and 22, 1830, in the Louvre courtyard trying to induce the artillerymen to deliver the cannon to the mob. Took part in nearly all riots and disturbances in Paris.

"On May 9, 1831, at the Republican banquet at Vandanges de Bourgogne, he drank the toast 'To Louis Philippe' with a dagger in his hand. Spent until June 15 in preventive imprisonment at Sainte-Pélagie. On June 15 was acquitted by the Court of Assizes, where he most violently attacked the government. Claimed that his toast was 'To Louis Philippe, if he betrays' and that he wanted either to kill the King himself, or to incite others to such an action in case the King should betray. Claimed that the King is likely to betray if he has not done so already.

"Character: In his speeches either very calm and ironical, or very passionate and violent. Seems to be a mathematical genius although not recognized by professional mathematicians. Lectured on mathematics in Caillot's bookshop, rue de Sorbonne 5. Did not use the lectures for Republican propaganda. Is one of the fiercest Republicans. Very courageous, extreme, and fanatic. No association with women. Can be very dangerous because of his daring. May have great influence upon people because of his fanaticism and ability to coin slogans. Easy for our men to approach because he generally trusts people and is inexperienced in worldly affairs."

M. Lavoyer finished and put the two pages neatly inside the dossier.

M. Gisquet said softly, as though to himself, "Very well done, very well done." Then drumming with his fingers upon the table, he said, "He looks like one of your first customers."

M. Lavoyer answered in a monotonous voice, but with an overtone of tender reverie, "Yes. He will be one of my first customers."

VII

SAINTE-PÉLAGIE

1 : July 14, 1831

It was forty-two years since the people of Paris had stormed the Bastille and impaled the heads of Delaunay and Flesselles on pikes. It was the first anniversary of that great Revolution to be celebrated since the new Revolution. Would the people commemorate their glorious past by renewing their eternal fight? The police prepared and made preventive arrests just before July 14. Raspail, among others, had been imprisoned in Sainte-Pélagie. He had written pamphlets arousing the people against their King.

Galois and Duchâtelet were ready to play their part on that July day. They were to lead the Republicans from the Champs Élysées to the Place de Grève where, forty-two years before, the people had massacred the defenders of the Bastille. Here, on the Place de Grève, they were to plant trees of liberty in memory of that liberty which always seemed near enough to be won today and far enough to be fought for again tomorrow.

At noon Galois and Duchâtelet, both dressed in the uniforms of the disbanded Artillery Guard, crossed the Pont Neuf, leading fifty Republicans. The bridge was densely flanked on both sides by gendarmes, police commissioners, and spies.

When the column entered the bridge, the police were watching the procession—apparently with indifference. But when the group reached the middle of the bridge, Galois saw a small detachment of gendarmes blocking the exit to the left bank of the river. The detachment stood at ease. It did not actively interfere with the traffic, but by its presence forced it to slow down, to split into two narrow streams, thus creating a congestion. Faced with this static, uniformed obstacle, Galoic and Duchâtelet turned to their right. The column of Republicans which they led spread in length and narrowed in width so as to pass through the free space between the detachment of gendarmes in the middle of the bridge and those stationed along the railings.

The commissioner gave a sign. Four of the gendarmes standing at the railings sprang forward into the free space between Galois, Duchâtelet, and their men. Two of them attacked Galois from behind, gripping the military collar of his uniform and immobilizing his hands. Then they pushed him expertly towards the railings and the Commissioner of Police. The two other gendarmes did the same to Duchâtelet. Simultaneously the uniformed detachment, which until then had stool calmly, charged into the column of Republicans, which dispersed immediately. All this was done quickly and skillfully. No one resisted.

The Commissioner of Police turned to Galois and Duchâtelet.

"You are Galois and you are Duchâtelet. I have warrants for your arrest."

Then to the four policemen:

"Take them to the dépôt."

Galois knew exactly what to expect. It was only two months since his first visit to this establishment. Today even the clerk in the long room became interested when he listed Galois' possessions: One loaded musket, one loaded pistol, one dagger.

2 : July 15, 1831

According to the law, every prisoner had to appear before the judge within twenty-four hours after imprisonment. Some twenty minutes before this time limit, two guards entered Galois' *pistole*. They carefully examined the walls and one of them remarked to the other, "No, nothing here." Then he showed handcuffs to Galois, swinging them so that they gave off a subdued metallic sound, and said, "If you promise that you will not try to escape, we won't put these on you. Do you promise?"

Galois nodded.

They took him out, crossed the street, entered another building, climbed a winding stone staircase, and deposited Galois in a room, where, at a large table and in a comfortable chair, sat a big man with a fat, oily face marked by signs of smallpox. Duchâtelet stood before him replying to his questions while the secretary made quick notes.

"It seems that you were not satisfied by the charge of plotting against the security of the state, but yesterday you committed a new crime in your cell."

Galois wondered what crime could be committed in a cell beyond killing bugs or misusing the *gogueneau.*

"On the walls of your *pistole* you drew a head and a guillotine, and you wrote underneath 'Philippe will yield his head on your altar, oh Liberty!' Is this true?"

"I did not draw a head; I drew a pear."

"You drew a pear. Then you will have to thank your Republican friends for making it abundantly clear that to them a pear represents the King's head. Why did you do it? What did you mean by it?"

"I meant what I said—neither more nor less."

"All right, M. Duchâtelet. I shall tell you why you did it. You thought we would be too stupid to see through your scheme."

His voice and tone were sweetness itself.

"You have studied law, haven't you, M. Duchâtelet? You were afraid that we might accuse you of a small offense, of wearing the uniform of the Artillery Guard. Of this small offense you can be convicted without a jury, just by a judge. This one judge could give you six months for it. So now you commit a great offense, you abuse the King. For this you think you will get a jury trial. Of course, you think, no one will care about the small offense of wearing an artillery uniform illegally. And a jury trial is just what you want. It makes you a hero and it gives you a great chance to make speeches and perhaps to be acquitted. Lately the juries have been very mild to Republicans. What about it, M. Duchâtelet? Am I right?"

"As you said, sir, I have studied law, so I know that I do not need to answer your question."

"Of course you don't need to. But you have made one mistake, M. Duchâtelet. You forgot that you can be tried for wearing a uniform by a judge, and then again for your subversive drawings by a jury. So you have not avoided any danger, you have run into two instead of one."

He looked at Duchâtelet benevolently.

"And in the meantime you will await your trial at La Force. It is, of course, merely a preventive arrest."

He still smiled, signed some papers and dismissed Duchâtelet.

Galois' turn came. On the table he saw his musket, pistol, and dagger.

The judge opened a dossier with many papers in it. Galois looked with pride at its thickness.

The judge asked many questions: about his parents, his brother and sister, about Louis-le-Grand and the Normal School. He checked the answers, looking at the papers before him.

Then he pointed toward the exhibition of arms on the table and asked, "Why did you carry all this?"

"To defend myself and to attack."

"To defend yourself against whom?"

"Those who might attack me."

"Whom did you expect would attack you?"

"Those who always attack the people."

"What do you mean by that?"

"I mean what I said."

"Whom would you defend with these weapons?"

"The people, if they were attacked, and myself."

Galois felt very tired. There was no fire, no irony in his answers. Their form seemed automatically to follow a preconceived pattern, as rigid as a mathematical formula.

"For this purpose would you have used not only the musket, the pistol, but also the dagger?"

"If necessary, yes."

"Don't you, an intelligent young man, think that the use of a dagger is much more barbaric and cruel than that of muskets and pistols?"

"I think it is cowardice and stupidity not to use any weapon which can be effective in an emergency."

"Do you, therefore, agree that your actions were intended against the security of the state?"

"No. A state in which such actions are necessary and possible has no security."

"But you were prepared to use your weapons?"

"Isn't it obvious?"

"Yes, it is quite obvious. Now tell me, why did you dress in the uniform of an artilleryman?"

"This is not the uniform of an artilleryman any longer."

"Do you deny that you are dressed in the uniform of an artilleryman?"

"The Artillery Guard was dissolved and therefore the uniform ceased to be the uniform of an artilleryman."

"That is a silly answer, prisoner Galois. The Artillery Guard was dissolved and this means that no one may wear its uniform after the day of its dissolution."

"I don't see the connection."

"You don't, but the judge may. In the meantime, you may meditate on this logical point during your preventive custody in Sainte-Pélagie."

The judge smiled sweetly.

The same day the *panier à salade* brought Galois to Sainte-Pélagie. He listened apathetically to the loud greetings.

"Here is Galois, our great scientist. Welcome, welcome again."

"I knew you would not stay away from us for long."

"He loves us! He had to come back to us."

He revived somewhat when he caught a glimpse of blond hair and a face he knew well. He rushed to Raspail. They shook hands, then both said almost simultaneously, "I'm glad you're here, old man," and laughed at their own stupidity.

3 : July 25, 1831

On this day Raspail wrote to a woman friend:

A new canteen has just been opened at Sainte-Pélagie for the high-class people; a prisoner has charge of it without a license, without permission, and without hindrance. You are served there as you are in a restaurant or a café. You can find everything there that the regulations do not allow in the official canteen; coffee and liquor flow freely; forbidden brandy comes in through the iron gate in a pair of boots which a woman brings back every day for repeated resolings. The jailer, who accompanies Madame to the iron gate, takes the odor of brandy to be the odor of Hungarian leather. And then, too, how could one have the slightest doubt as to the veracity of a pretty woman who, each morning before going to the prison, visits M. Parisot, Chief of the Division of Prisons in the Department of the Seine?

That canteen drives me to despair. For our high-class drinkers end up by dragging there the noblest souls among our young companions.

"Come, come, my poor Evariste! You must become one of us! Take this little glass as a trial. You are not a man until you have had women and good wine."

To refuse this challenge would be an act of cowardice. And our poor Evariste has in his frail body so much courage that he would give his life for the hundredth part of the smallest good deed. He grasps the little glass like Socrates courageously taking the hemlock; he swallows it at one gulp, not without blinking and making a wry face. A second glass is not harder to empty than the first; and then the third. The beginner loses his equilibrium. Triumph! Victory! Homage to the Bacchus of the jail. You have intoxicated a candid soul, which holds wine in horror!

Mercy, mercy on this child so puny and so brave, on whose brow three years of study have deeply engraved sixty years of the most learned meditations. In the name of science and of virtue let him live! In three years he will be the great scholar, Evariste Galois.

But the police do not believe that scholars with pure souls exist. How cheap would the secretaries and heads of departments appear, those men who were or may become professors, and are devout or liberal as ordered, if the seed of this young scholar spread on the soil of our unfortunate country?

I do not doubt, madame, that Galois would inspire in you a venerable interest. Oh, if he had a sister such as you, how he would forget his mother!

On the thirteenth, this child was told that the following day all devoted men were getting ready to defend their principles with weapons in their hands. He replied, "We shall be there, my comrade and I; we are going to grow several inches taller," and each of them wore a complete artilleryman's uniform with arms and equipment. They stuffed themselves with bullets, powder, pistols of all kinds. And I assure you that, if Galois had returned from the battle, he would not have brought back an ounce of all this ammunition. I assure you that some of Galois' accomplices feared his presence on July 14 as much as the police itself; they felt more at their ease, I am sure, when they learned that he had been arrested. What is not to be feared nowadays, in one way or another, from a virtuous man who plans all his actions with mathematical precision?

Galois was wandering one day in the yard of the prison like a man who belongs to this earth only because of his body, and who lives only by his thoughts.

Our canteen bullies cried out to him from the window, "Heh! You old man of twenty. You haven't even the strength to drink, you are afraid of drinking."

He went upstairs, walked straight toward danger, and emptied a bottle at one draught. Then he threw it at the head of his impertinent aggressor. What divine justice if he had killed him on the spot! It was a bottle of brandy!

Galois went back down the steps, straight and steady on his legs; the liquor had not yet passed his gullet. But what misfortune once it reached his stomach! I never saw such a prompt upheaval in the habits of a poor little creature! He straightened up with all the pride in his soul; one would have thought he was growing taller and that he was going to use up in an hour all the abundance of strength which nature might lavish on him for the next twenty years.

He clung to my arm like a climbing plant that looks for support and said, "How I like you, and at this moment more than ever. You do not get drunk, you are serious and a friend of the poor. But what is happening in my body? I carry two men inside me, and unfortunately I can guess which is going to overcome the other. I am too impatient to get to the goal. The passions of my age are all imbued with impatience. Even virtue has that vice with us. See here! I do not like liquor. And at a word I drink it, holding my nose, and I get drunk. I do not like women. It seems to me that I could love only a Tarpeia or a Graccha.

"Do you know, my friend, what I lack? I am confiding it only to you. It is someone whom I can love and love with my whole heart. I have lost my father and no one has ever replaced him. Do you hear me? Oh, how good you are to me not to laugh at me as these base actors of the lowest melodrama would do. I shudder when I hear their voices! What a filthy hole we are in! If someone would only take me out of here!"

You may well imagine that no matter how touching the language of this pure soul was to me, I was only looking for a favorable opportunity to put an end to it. I pulled him gently toward the stairway and I made him climb up to his room. At that moment the bell rang for locking up. My cellmates respected his misfortune and, without much trouble, the turnkeys agreed that only the door to the stairs should be locked and that the door between our patient's cell and ours should be left open. We laid him out on one of our beds. But the fever of intoxication tormented our unhappy friend. He would fall back senseless only to raise himself with new exaltation, and he foretold sublime things which a certain reticence often rendered ridiculous.

"You despise me, you who are my friend! You are right. But I, the culprit, must kill myself."

And he would have done it, if we had not flung ourselves on him, for he had a weapon in his hands. God finally took pity on his sufferings. His intoxication burst forth in a fit of vomiting which inundated the room in torrents. And the unfortunate patient fell asleep, relying on his good comrades to repair the damage. The floor was flooded. We put on wooden shoes and we pounded with heavy blows on the door; it remained deaf to our cries. One of us, more courageous than the others, began to scoop up in the hollow of his hands and to throw away in the *gogueneau* the streams which followed the irregular slopes in all the directions of the room. We mopped up the rest as best as we could with the utensils we had. We put the filth in the patient's room and the patient in ours. And the following morning Science and Liberty had recovered their worthiest adept. Our young comrade had rewon our esteem and the sick man our forgetfulness. Unfortunate child! In order to save himself from the snares which await him in all the events of his fine career, he only needs to have a little distrust. But nature does not bestow this quality. We can only obtain it, to our own detriment, from our dealings with men. Oh, society! Here is the dilemma you impose: either to be the dupe of evil or lose all belief in goodness! But there are some beings whose guardian angels remove them from this earth the instant their youthful eyes perceive the end of this inexorable sophism of our false institutions.

4: August 2, 1831

On this day Raspail wrote to his friend:

Since I last wrote you, madame, important events have taken place in this little corner where the law confines us. We have celebrated the anniversary of our three glorious days. It seemed for a time that the anniversary would change into three days of mourning.

On July 27, the prisoners of Sainte-Pélagie were invited to Mass in memory of the July dead. If we had destroyed the catafalque on that saintly day, they might have been able to punish the offense as sacrilege. And this word would have taken hold, for Paris still honors these illustrious dead as saints. It would have applauded the public prosecutor who by means of this odious epithet could claim to have defeated an attempt to profane the memory of its bravest children.

As we came out of Mass, two or three voices from the crowd suddenly shouted that the pitiful catafalque must be destroyed; that it was an insult to the memory of our glorious Revolution. For the past year, our government has so often insulted this memory, that it is hardly able to say a word about it now, which does not seem to be an insult. And without two or three sound heads scattered through the crowd, the motion would have been carried out all the more easily since the turnkeys, for reasons known to themselves, cared little about watching. Prudence lies not in fighting an evil thought directly, but in substituting for it a new and inoffensive thought. This was done with such skill that the instigators could still pride themselves on the wisdom of the accomplished deed. Instead of the catafalque being shattered, it was carried into the yard so that it might stay there three days, an object of mourning to impose respectful silence in the name of the most pious sorrow and the saintliest regrets.

The twenty-ninth arrived without the slightest incident. The bell to announce locking-up time rang, and the expected rebellion did not come.

Bang! A shot was fired just when the locking up was finished. We heard cries, "Help! Murder!"

Repeated blows shook the doors of several cells. The jingling of keys indicated to us that the turnkeys had redescended accompanied by two or three prisoners who were exasperated by pain and indignation. Following this, profound silence all night long. You realize, Madame, what thoughts must have passed through the minds of all of us who had twelve hours to wait before being able to communicate with each other and to find the source of such an unusual event.

At the opening of the doors, each cell and each enclosure poured forth its quota of prisoners into the yard, from which the torrent, in spite of the efforts of the turnkeys, invaded the clerk's office and that of the superintendent. This official, horrified and losing his head, would have been suffocated in the arms of the exasperated prisoners without the obliging intervention of the general inspector and other counsellors of the prefecture who appeared in the nick of time.

"Do you want to butcher us then, defenseless and one by one?"

The prisoners cried out this question with that unanimity of expression which a crowd shows when moved by the same passion.

The high employees of the police replied with that unmoved official tone which cloaks the executioner when he invites the culprit to put his head in the guillotine.

"No, such is not the intention of the administration. We are not thinking of slaughtering you."

"Well, then, where are the three prisoners who are missing?"

"In the dungeon."

"In the dungeon! What have they done to deserve the dungeon?"

"They vociferated and complained insolently!"

"Insolently! How can one be insolent to people like you? What did they complain about?"

"One of them said he had received a shot in the face, and the other confirmed the fact."

"Is this true?"

"Certainly."

"And do you know who is guilty?"

"We suspect him."

"And he isn't in the dungeon?"

"Justice is investigating."

"Why don't you skip that phrase at which even the grocers are beginning to smile. Justice investigates! Indeed! Don't speak to us of justice. You threw our friends into the dungeons without it, since it is busy investigating. Why hasn't the person guilty of this atrocious crime, atrocious as to content and as to form, been seized by you?"

"We still have doubts."

A voice of a man of the people cried out with paralyzing force, "You lie in your throats, you informers. I saw everything and I know everything. And if I didn't ask to go down last night, it is because I knew that, with a Philistine like this cunning superintendent, it was best to keep quiet so as to be a witness on the following day. Do you see him getting pale and losing his poise? This protector of murderers knows that I am going to tell the truth."

The superintendent stammered, "I fear nothing."

"No, but you're trembling. You mean that you don't repent anything and that you had orders. Here are the facts. I am one of those in the room which is under the roof of the bathing pavilion. We were quietly going to bed. The one whose bed is between two casements had his face at the window while undressing and he was humming a tune.

"At that moment a shot was fired from the garret opposite. We thought our comrade was dead, but he was only unconscious. Not knowing from where the shot came, nor how serious the wound was, we called for help. For in a room thus open on all sides by six windows, each better-aimed shot would have ended by striking down its man. You know what help they brought us. But you must know too, the person guilty of this trap."

"We have doubts about that."

"I am going to make you certain. The man who fired the shot rooms in the garret. He is still there. Have him sent for."

"We have no right to do that."

"Why not? He is a turnkey who yesterday and the day before yesterday was still guarding our door."

"One of our turnkeys?"

The exasperated crowd shouted, "Yes, one of your turnkeys. He rooms in the garret on the rue du Puits-de-l'Ermite. You know it very well. And if you want to make sure, choose any one of us, he'll take you there immediately."

"We have no orders in this matter."

"What? You have no orders to seize the guilty man? And you throw into the dungeon both the victim of this shameful trap and the witnesses of it? It may sound insolent to say that the administration pays turnkeys to murder prisoners. But what if this insolent statement is true? And I bear witness that no other insolence has come from those who were thrown in the dungeon. This young Galois doesn't raise his voice, you know it well; he remains as cold as mathematics when he talks to you."

"Galois in the dungeon! Oh, the bastards! They have a grudge against our little scholar."

"Of course they have a grudge against him. They trick him like vipers. They entice him into all sorts of imaginable traps. And then, too, they want an uprising."

"They will have it and they will be able to slaughter all of us. It is better to die together than to let ourselves be butchered one by one like pigeons. Down with the police spies! Get out of here, you murderers! The prison is ours! We shall make it our stronghold! Forward!"

At this cry, still ringing in my ears, you should have seen, madame, how all these representatives of authority took to their heels, and with what speed those doors, formerly so heavy, swung on their hinges all at the same time.

The prisoners closed the doors behind the turnkeys. The tables and desks were used to barricade doors and windows. Iron chains were twisted around the grating in the yard to form a lock, for which only a file could be the key. Now that an uprising was proclaimed by the political prisoners, the small urchins recovered their liberty. They could have escaped from the prison, but they did not think of it. They enrolled, they said, in the service of liberty. The quickness with which these little mice began to gnaw the iron links of the mousetrap which the grown-ups had such trouble in shaking, is still beyond my comprehension. The iron bars twisted and broke between their small fingers as readily as glass tubes. The railings of the stairs disappeared noiselessly from the ground floor to the fifth story, melting before our eyes as in a huge crucible. And fifteen minutes later there were no traces of them left. Where did all the iron go with which the prison had been bristling a quarter of an hour before? I will let you guess. We rummaged in all the corners of the building, but we found none of it though nothing had been taken out. And these little monkeys listened with the greatest indifference to the various questions which the prisoners asked each other regarding this matter.

Be that as it may, the day was spent in preparation for defense in the most peaceful atmosphere. The prison was without supervision and it was never quieter. Never had order reigned so completely in this assembly of men, whom our excellent law pursues as partisans of disorder. It is a curious thing to see how men live in peace as soon as they no longer have masters. The masters maintain just the opposite, but they are lying, Madame, you may be sure.

From morning till night we remained the owners of the building, of the offices, of the registry, of the apartments, of the canteen. When they take inventory again, I can assure you that not a single little glass of the canteenkeeper will be missing. We had the keys, even the huge key of the door which communicates with the detention room. The children got hold of it by tripping the terrified turnkey as he fled.

Without a doubt the administration, gathered in council, was not waiting with folded arms for events to develop. But certainly no appeal was made to us. And no battalion of the National Guard was ordered to besiege us.

A spirit of revolution reigned among the prisoners. There was great bravery and devotion to the common cause. The old officer among us who imitates Napoleon strode around the yard, staring fixedly in front of him, hands behind his back. You can't imagine the pleasure one feels in getting rid of evil, and how free and easy our hearts become as soon as we have reason to believe ourselves surrounded by friends.

I will confess my childishness: never did resistance to injustice present itself to me in a more attractive light. And when in the evening at dusk, the pavement of the neighboring streets began to echo with the galloping of cavalry which emerged in squadrons, and when over the noise of the butts of infantry rifles the words, "Halt and rifles to the ground!" were wafted to us, I understood the sublime impatience of Job's quivering charger as he catches the scent of battle with his burning nostrils. Oh, who will ever give me the opportunity of fighting for a good cause, side by side with men who I know for sure are not spies?

Suddenly the great outer door was opened and a host of councilors of the Prefecture appeared at the gate to enter into negotiations.

The prisoners asked them, "Is the Prefect here?"

"He won't come until very late. He is at a court ball."

"At a court ball, at a time like this! How strange they are at court, to dance while so many wretches are suffering and in revolt! Well, then, go and remind him that his place is not there, and tell him that the prisoners wish to speak to him."

"We have the same authority as he, and we represent him here; return to order, or else—"

"Or else!"

At these words I saw the crowd of official hats and tricolor scarves step back with one jump into the street. Authority itself, deeply compromised, cried out in the act of coughing, sneezing and blowing its nose, "Guards, guards, help!"

And the guards, in groups, charged with fixed bayonets up to the gate.

The prisoners shouted, "That's a fine place to break your neck. All you have to do is to stay quietly where you are. Do not fear. We will start only if you do."

And once again the prison assumed the air of calm that it had had all day, just as if an army of besiegers were not waiting at the door for the signal to start the assault. It was one of those beautiful summer nights, when you like to be in the shade as though it were day. Our little urchins, stirred up by what had happened, began to sing hymns with rhythm and harmony. Their music teacher beat out the time. Unfortunate little outcasts of society with their swan's voices, they sang and saluted the night, which came to them in their jail on the wings of a light breeze together with one more star, a falling and fugitive star, which flashes through the present and loses itself in the future—Liberty!

We lent such an attentive ear to these charming childrens' voices, that we forgot the strange and difficult circumstances that held us assembled at such a late hour outside our cells and free from our turnkeys' watch. Nevertheless, those of us who were less accessible to the charms of musical language could have noticed, prowling around the group, certain familiar figures which had always aroused in us the gravest suspicions. With a little bit more of that distrust which the wise man should always keep in times of revolution, we would have seen these shadows directing their steps toward the doors of the grating to shake its chains.

Suddenly a cry, "Everyone for himself!" interrupted the nocturne. The iron gates which we had barricaded with so much skill opened as if by magic. The infantry came into the yard with bayonets fixed. Our blusterers fled to the higher stories. Even Napoleon's hat turned completely around for the first time in his glorious career. The enemy was sweeping the yard practically at a running pace, when suddenly we saw it draw back, like a traveler, fearless till then, who has just stepped on the tail of a snake.

The road to victory was blocked by a double row of small urchins, the youngest in the front row, the tallest in the second, and all armed with those same iron bars for which we had looked in vain all day. These little devils had held them in their trousers until the favorable moment. And not one of their movements, not one of their provocations, not one of their capers had been hampered in the slightest by the stiff iron bars which they carried under their clothes. We saw them brandishing their weapons with a vigor which left no doubt that blood would have flowed as on a field of battle had not the police commissioner prudently decided to capitulate and had not the prisoners in whom these indomitable Lilliputians had placed their confidence accepted the offer of capitulation.

The first row was already preparing itself to throw its iron bars at the guards' faces, and then to make its way between the legs of those still standing and to throw them on their backs. Meanwhile, the second row would have fought on by using its weapons as swords to complete the youngsters' work.

Do not laugh at these claims. What I am telling you is as serious as it seemed, and no one on the field of battle felt like laughing. But even if the revolt were overthrown in the prison yard downstairs, it still had a chance in the upper stories. With the door to the detention room once opened by the key we possessed, a fire made of all the ticks and mattresses of the prison would have protected the retreat of the insurgents.

The resulting disorder would have made our escape easy, and the disappointed victor would only have won several blackened walls.

Capitulation was honorable for the youngsters. Not a hair of their heads was harmed. Complete amends were made to the prisoners, to whom assurance was given that their comrades would be returned to them the following morning at daybreak. The lesson cost the administration more than it had estimated. Twenty-thousand francs will not indemnify it for the havoc which it provoked in demanding a riot. It will doubtless take revenge. But in the end power was left in the hands of the just, as against the brute force and Machiavellianism of the crafty. Order reigns at Sainte-Pélagie! The prisoners know how to maintain it despite the jail. The prison has become a topsy-turvy world; the people governing it were those jailed by the administration which came to power through the July Revolution.

5 : 1831

Galois spent July 30 in a dungeon, a dark, small cell closed with heavy doors. He lay on the hard bed and tried to recall objectively the events that had brought him there.

Four prisoners had been in his cell. It was a good, comfortable cell with six windows, very convenient for shooting practice from the garret across the rue du Puits-de-l'Ermite. They had been undressing when he heard a loud shot and saw the wall opposite the window pierced by a bullet that passed between him and his comrade, who then fainted. Galois was sure that it just missed his head.

The shouts of the prisoners brought three turnkeys and the superintendent, himself, in person. That important official did not bother to look at the prisoner who had fainted. He just stared at Galois, who could swear that he saw the following sequence of emotions on the superintendent's face: astonishment, disappointment, fury, wild rage.

Then the high official raised his fist and yelled at Galois, "You were shooting at your comrade. You want to assassinate someone, you don't care whom. I'll show you. You will. . . ."

Galois had stood quietly, too enraged not to be calm. He would have acted differently if he had had his weapons. One of the prisoners lay on the floor unconscious and another sat silently on the bed.

With courage the third jumped at the superintendent and cried out, "You are trying to kill us and you accuse others of doing it."

The unconscious man on the floor now gave his first sign of life by kicking one of the turnkeys with his feet.

The superintendent pointed to Galois, then to the prostrate man on the floor, then to the prisoner who had jumped at him, and commanded, "Take these three to the dungeon."

They had left one prisoner in the cell, the one who sat silently on his bed. Galois smiled with satisfaction. Here they had made a mistake. He knew that this prisoner who acted so calmly was courageous and prudent. It was not hard to guess why he had kept quiet.

"At whom was the bullet aimed?"

He must talk about it with Raspail. He must tell him what he suspected: that the bullet was aimed at him.

His thoughts were interrupted by a loud banging on the door.

"A revolution has broken out at Sainte-Pélagie. We are in possession of the prison. Wait. We shall open your door, and you will be free."

They shouted back and forth the story of the revolt until they ran in search of greater excitement, promising to come back and break down the door. But they did not come back.

He thought how a year before he had been imprisoned in the Preparatory School while the victorious Revolution swept over Paris. Today he was in a dungeon while a victorious revolution swept over Sainte-Pélagie.

"What progress—and all achieved in one year! From school to Sainte-Pélagie. My role is the same, always the same, to remain imprisoned, inactive. But at least today I am one of those who provoked the revolution. What an achievement, what a tremendous achievement! The prisoners revolted to save me from the dungeon."

But in the heat of their fight the prisoners forgot those for whose liberation they had begun their battle. It was not until a day later that Galois and his two comrades were freed from their solitary cells by the lawful prison authorities.

He often returned to the thought of the bullet which had been shot into his cell. He was ashamed to tell Raspail about his fears and suspicions. Was he really important enough to be the target of a bullet designed especially for him by an intricate plot? It sounded too fantastic and too conceited. But then he remembered how M. Dupont, his counselor, had warned him. Even if his suspicions were justified, how could he prove anything? Was it not obvious that the man who fired the shot would say that he did it accidentally while cleaning his gun? And besides, with every passing day, he felt more tired and more indifferent to the world around him.

In October, Galois received a letter for which he had been waiting over two years, a letter bearing the seal of the Institute. It had finally found its way to Sainte-Pélagie. The envelope was big and fat. He opened it, pretending to be calm and disinterested. It contained his paper and a letter from the Secretary of the Academy:

Dear M. Galois,

Your paper was sent to M. Poisson to referee. He has returned it with his report from which we quote:

"We have made every effort to understand M. Galois' proofs. His argument is neither sufficiently clear nor sufficiently developed to allow us to judge its rigor; it is not even possible for us to give an idea of this paper.

"The author claims that the propositions contained in his manuscript are a part of a general theory which has rich applications. Often different parts of a theory clarify each other and can more easily be understood when taken together than when taken in isolation. One should rather wait to form a definite opinion, therefore, until the author publishes a more complete account of his work."

For this reason we are returning your manuscript in the hope that you will find M. Poisson's remarks useful in your future work.

The letter was signed by François Arago, Secretary of the Academy.

Galois went from the yard to his cell. There was no one else there. He read the letter again, his mouth curved in a grimace of loathing. He tore the letter into four pieces, tore them again into smaller and smaller pieces. Then he squeezed them tightly in one fist, opened the *gogueneau* with the other hand and, holding his breath, let them drop inside. Then he covered the piece of furniture, sprang away from it toward the window, and breathed deeply.

He looked through his rejected paper. Yes, it was the same one which he had sent to the academy ten months before. On the top of the first page someone had written, "M. Lacroix, M. Poisson, commissaires." He looked at a comment penciled by Poisson on the margin of the third page: "The proof of this lemma is not sufficient. But it is true according to Lagrange's paper No. 100, Berlin, 1775."

His head whirled with disordered thoughts until they crystallized into contempt and hatred.

"They understood nothing! And all that I asked of them was to read these few pages carefully. Yes, they are difficult. But if I had expanded them into a whole book they would have said, these foolish academicians, that it is too long and that they do not have time to read it. These little men. I'll show them. I must make my results known. Perhaps someone will read; perhaps someone will understand.

"And I must tell the world how these men treated me. Let future generations judge between Poisson and Galois, between the Academy and me. They must be punished for what they have done to me. Here at Sainte-Pélagie where I am being punished by a cruel and stupid regime I shall punish all these vain men, smug and satisfied because they have pushed mathematics such a small step forward that everyone can measure and understand what they have done."

An idea occurred to him and he became excited, elaborating it into a concrete plan. Here he held a manuscript which the Academy had rejected. Among his manuscripts was a second paper on the theory of equations, almost finished. It would be easy to rewrite and complete it, since all the results were clear in his mind. He would make known these two papers that gave the solution of the central problem of algebra and contained results of the most fundamental nature. He knew their importance! He would print them himself in a small booklet! It should not cost much. Only the two papers and the introduction! The introduction! This introduction must be such that everyone, even Cauchy, Poisson, and the examiners of the Polytechnical School would read it, though with very little pleasure.

To whom should he send this booklet? He imagined a pile of copies lying on his table; they would look like the pamphlets of the Society of the Friends of the People, but the front pages must be more attractive and the covers should be red. He would send them all over the world. He would make a list. He wrote down the names of Gauss, Jacobi, the names of prominent French mathematicians, of Lacroix, Poisson, Cauchy. Let them see that they could not discourage Evariste Galois. Let them read the introduction, let them blush and be ashamed of the shameless things done. Then he wrote the names of his teachers: Vernier, Richard, Leroy. He would send some copies to institutions. Of course! He must not forget the Normal School. M. Guigniault ought to have Galois' booklet in his library. Let him read the introduction. And of course he would send a copy with his compliments to the Polytechnical School. And a copy to the institute! He liked the list. Now the introduction. In a few days everything could be ready for print. He would ask Chevalier to arrange the formalities. His friend would be happy to do it.

Evariste wrote the title:

TWO PAPERS ON PURE ANALYSIS

by

E. Galois.

PREFACE

He thought, "How shall I begin? Has anyone else at twenty written a paper as important as mine? True, when Gauss wrote his *Disquisi-*

tiones he was only slightly older than I am. But how he was treated —and how am I treated? I remember the first page of *Disquisitiones,* where he thanks his patron, the Duke of Brunswick, in large print. Gauss had his benefactor. But who helped me? I don't want help. I wish I had enemies who would oppose my work, discuss it, try to refute it. No! It is much worse! Indifference, emptiness, silence are my fate. No one understands and no one tries to understand my work."

Evariste wrote the introduction. Rage and scorn drove his hand and pen. At tremendous speed he poured out words and sentences, crossing out some, substituting others, covering the page quickly with irregular, dense, and disorderly writing. Thus he wrote the preface:

Firstly, you will notice that the second page of this work is not encumbered by surnames, Christian names, titles, honors and the eulogy of some niggardly prince whose purse would have opened at the smoke of incense, threatening to close when the incense holder was empty. Neither will you see in letters three times as high as your head, homage respectfully paid to some high-ranking personality in science, or to some wise patron, a thing thought to be indispensable (I was going to say inevitable) for anyone wishing to write at twenty. I tell no one that I owe everything of value in my work to his advice or to his encouragement. I do not say it because it would be a lie. I swear that it is not thanks that I owe to men important in the world or important in science.

I owe to the men important in science the fact that the first of the two papers in this work is appearing so late. I owe to the men important in the world the fact that I wrote the whole thing in prison, an abode hardly suited for meditation, and where I have often found myself dumbfounded at my own listlessness in keeping my mouth shut in the face of my stupid, ignorant, spiteful critics. The why's and wherefore's of my stay in prison have nothing to do with the subject at hand. (The author is a Republican and a member of the Society of the Friends of the People. He has shown by a gesture that regicide may be desirable.) But I must tell how manuscripts go astray most frequently in the portfolios of the gentlemen from the Institute, although I can hardly conceive of such absent-mindedness on the part of those who already have the death of Abel on their consciences. It will be sufficient for me, since I do not wish to compare myself with that celebrated mathematician, to say that my paper on the theory of equations was sent to the Academy in February, 1830 (in less complete form in 1829), that I never again heard anything of these manuscripts, and that it was impossible for me to get them back. Perhaps I have said too much, but I wanted to explain to the reader why it would have been absolutely impossible for me to embellish or disfigure my work by a dedication.

Secondly, the two treatises are short. There is at least as much French in them as algebra. To these charges I plead guilty. Indeed, I could have increased the number of equations by substituting in succession all the letters of the alphabet in each equation, numbering them in order. This would have multiplied indefinitely the number of equations, if you reflect that after the Latin alphabet there is still the Greek alphabet and, if these are exhausted, nothing stops us from using Arabic letters and if need be, Chinese! It would have been so easy to change each phrase ten times, being careful to precede each change by a solemn word *theorem;* to arrive at results by *our analysis* in the good old Euclidean fashion; to precede and follow each proposition by a formidable lineup of special examples. And from so many methods I wasn't able to choose a single one!

Thirdly, it must be admitted that a master's eye saw the first of the two papers printed here. An extract sent in 1831 to the Academy was submitted for M. Poisson's inspection, who said that he understood nothing of it. To my conceited eyes this proves simply that M. Poisson either did not want to, or could not, understand my work. But it will certainly prove to the eyes of the public that my book is meaningless.

Therefore, I have plenty of reasons to believe that the scientific world will greet my work, which I am submitting to the public, with a smile of compassion; that the most indulgent will accuse me of blunder; and that for some time I shall be compared to those untiring men who each year square the circle anew. I shall especially have to bear the wild laughter of the Polytechnical School examiners who, having monopolized the printing of mathematical textbooks, will raise their eyebrows because a young man twice refused by them has the pretension to write not a textbook, but a treatise. (In passing: I am very much surprised that the examiners do not all occupy chairs in the Academy, since their places are certainly not in posterity.)

All the foregoing I have told to prove that it is knowingly that I expose myself to the derision of fools.

If, in spite of everything, I publish the fruit of my labors with such slight chance of being understood, it is in order that the friends whom I found in the world, before I was buried under lock and key, may know that I am still alive, it is perhaps also in the hope that my work may fall into the hands of men whom a stupid arrogance will not deter from reading it, and may direct them into the new road which, in my opinion, analysis must take.

It must be understood that I am speaking of pure analysis.

As new pages were covered with his quick, nervous writing, he felt a growing relief. This was his answer to Poisson's judgment on his work. Now hand and mind drove the pen. He regained calm and self-control, writing about mathematics, about analysis, about the importance of elegance and simplicity, and about the future development of algebra.

He had reached the concluding sentences of his preface. Now he felt high above the Academicians, secure and confident in himself when he wrote:

The general thesis which I advance can only be understood if one reads attentively that part of my work which is only an application of this general thesis. Not that the theoretical view preceded the applications! But I asked myself, after I had finished my work, why it is so strange and difficult for the reader. And I believe the reason for this is my tendency to avoid the use of formalisms and calculations; moreover, I recognize an insurmountable difficulty in working out such a general formalism in the subject matter which I treat.

One can easily surmise that while working on such a new subject, along such unusual lines, I have often encountered difficulties which I could not overcome. Thus in these two papers, and especially in the second, the reader will find the remark, "I don't know." I am aware that by writing this I am exposing myself to the laughter of fools. Unfortunately, hardly anyone realizes that the most precious and scholarly books are those in which the author states clearly what he does not know, because an author harms his reader most by concealing a difficulty.

When competition—that is, selfishness—no longer rules in science, when people associate with one another for study and not in order to send sealed packages to the Academies, they will be eager to publish even small results, as long as these are new, while adding, "I do not know the rest."

Evariste finished, relieved and tired. He looked at the blank walls and grated windows, then he wrote, "Sainte-Pélagie, October, 1831. Evariste Galois."

He had covered four pages. Scorn and anger vanished; only listlessness and apathy remained. His plan of printing the booklet now seemed to him childish and stupid. What did he care about Cauchy and Poisson and the examiners of the Polytechnical School? What did he care about anything in the world? But he did care about the moments when darkness is banished and new vistas opened. Here in the rotten hole called Sainte-Pélagie, these were the only moments worth living for.

6 : 1831

The machinery of justice slowly turned its heavy, rusty wheels. There was no escape for those who were caught, turned, broken, by the constant and uniform rotation. Sometimes the machinery seemed to stand still; sometimes it seemed to have lost its victim. But soon he was reminded that the system was not static; his turn would come.

Galois waited through July, August, and September, into the beginning of October, and still there was no trial. Over three months in preventive arrest! The last time he had been in Sainte-Pélagie only a month, but this time it was already three months—three months merely for wearing the uniform of the Artillery Guard. No, even he had underestimated the tyranny of the regime. Even if he were acquitted, then these three months spent in misery and despair could never be restored. If he were sentenced to two weeks, then he would still have to remain two weeks besides the three months of his preventive arrest. By calling it "preventive" they made it nonexistent in the eyes of the law. But it was as real and as horrid as any imprisonment served after sentence.

Evariste asked himself, "How many days longer? How many weeks longer? It depends on the trial. But the result of my trial is determined only up to an arbitrary constant. This constant of integration must be added to any verdict. It has been my privilege to find out how big this constant can be.

"They cannot accuse me of much. All I did was to wear the uniform of the Artillery Guard. Many others did the same. That was my great mistake. I did something too insignificant for a jury trial—something trivial, unimportant. Now they have me in their clutches quietly, discreetly. They don't even need a jury. Justice works better and more efficiently if it looks less majestic. I was acquitted when I incited the people against Louis Philippe. But I am and I will be punished for wearing the uniform of the Artillery Guard. I will be punished, because I am not in the power of twelve men but in the power of one man; one judge, as unimportant and as insignificant as the crime of which I am accused. This judge can easily be dismissed, and that is why he has to know how to take orders and please his masters."

Duchâtelet was acquitted for drawing the pear and offering it to the guillotine in the name of liberty. But not until the end of October were Duchâtelet and Galois tried for wearing the uniform of the Artillery Guard. They had violated Article 259 of the penal code. The trial was brief, the judge determined and not very talkative. He announced the verdict: Three months imprisonment for Duchâtelet and six months imprisonment for Galois.

The Orléanist press rejoiced. It had rejoiced once before when Galois, the great enemy of the King, was caught on July 14, and it rejoiced again when he was sentenced to six months after spending over three months in preventive arrest. Over nine months altogether! Why was Galois' sentence twice as severe as Duchâtelet's? The answer was simple, and the Orléanist press decided that the judge was wise and just. Galois deserved a sentence twice as severe as Duchâtelet. It was true that both were caught at the same time under the same circumstances. It was true that both wore the uniform of the dissolved Artillery Guard. It was true that each of them had a loaded pistol and a loaded musket.

But there was a very essential difference. Galois had a dagger while Duchâtelet had none. If you took the scale of justice, put on one side of the scale the uniform, the pistol, and the musket, and on the other side of the scale only the dagger, then the sensitive scale of justice would be in perfect equilibrium. The dagger was made heavy and important by the words "To Louis Philippe" which were once spoken when that same dagger was raised. It was therefore just that Galois, who carried with him twice the weight Duchâtelet did, if determined by the scale of justice, should receive a sentence twice as long—six months.

Thus the men of the bourgeoisie who loved Louis Philippe were happy to learn that the life of their King was safe for the next six months. They were only afraid that the Court of Appeals might change this sentence. Thus with relief they read to themselves, or aloud to their fat wives, or to their whole families, the report on the appeal trial as printed in the *Gazette des Tribunaux:*

Royal Court of Paris (*appeals before the magistrate*)
(M. Dehaussy presiding)
Hearing of December 3.
Trial of Messieurs Galois and Duchâtelet for usurping uniforms and carrying forbidden weapons.

The *Gazette des Tribunaux* gave an account of the trial which condemned two young men, Messieurs Galois and Duchâtalet, the former to six months, the latter to three months imprisonment for being arrested

on July 14 in the uniform of artillerymen of the National Guard, each of them carrying a loaded musket and a loaded pistol, to which M. Galois had added a dagger.

Both lodged an appeal against this sentence. M. Galois is the young man who was indicted in the month of June for proposing a certain toast at the banquet in the Vendanges de Bourgogne, but who was acquitted and immediately released. The public prosecutor also appealed for a more severe sentence.

Questioned by the presiding judge, M. Dehaussy, the accused both declared that they wanted to be present at the planting of the trees of liberty, and fearing that they would be insulted and attacked, they had armed themselves and thought that they had the right to wear once more their former uniforms of the artillery of the National Guard.

M. Chauvin, painter, former artilleryman, declared that after the dissolution of the artillery of the National Guard on January 1, 1831, those who belonged to it believed it their right to continue wearing its uniform.

M. Raspail, writer, now prisoner at Sainte-Pélagie, made a similar statement.

M. Bixio, a medical student, said, "In fact the artillerymen looked upon the dissolution of the Artillery Guard as illegal. I do not hesitate to wear the uniform every time I think it necessary for my safety, and, to my knowledge, the public prosecutor has never started proceedings against me."

M. Benoît, Superintendent of Police of the District of Saint-Victor, gave evidence that on the morning of July 14 he went to Galois' home with a warrant for arrest issued in his name. Because of the absence of the accused this warrant was not served.

Messieurs Ledru and Moulin, counsels for the accused, gave proof that these young men were members of the artillery of the National Guard and appealed to the custom which allows soldiers to retain their uniforms for a certain length of time after setting them aside.

M. Tarbe, the prosecutor, refuted the defense. He said that Messieurs Galois and Duchâtelet must be condemned for the double offense, firstly for wearing the uniforms, and secondly for carrying forbidden weapons.

M. Moulin was amazed to see the public prosecutor maintain his appeal for a more severe sentence as if the age of the two young men, the five months' imprisonment which they had already undergone, and their good faith, did not constitute extenuating circumstances.

The magistrates retired and after half an hour of deliberation rendered the following decision:

Considering that the illegal wearing of uniforms, of which Duchâtelet and Galois are guilty, is aggravated by a circumstance that they were both bearers of loaded muskets and pistols and that moreover Galois was the bearer of a dagger hidden in his clothing:

The law rejects the appeal and orders the sentence to receive its full and entire execution.

Thus when he reached the end, the bourgeois joyfully concluded that order, law, and justice reigned in France. He accepted a kiss on his hand from the children, nodded benevolently to his wife, took his tall black silk hat, his yellow gloves, his silver-headed cane, and proceeded in peace and happiness toward the Bourse.

7 : 1832

Galois tossed on his cot, reliving the details of the day's family scene. His older sister, Madame Chantelot, and his younger brother, Alfred, had visited him. His older sister had become more ladylike with every day of her marriage. When Evariste asked about his mother, tears came to her eyes.

She said affectedly, "I am glad that you mention Mother. She suffers very much. If not for your sake, then for her sake, please try to live a normal life when you leave these walls."

Yes. Those were her exact words. How many times had he heard them since his father died? To live a normal life for his sake, for his mother's sake, for all the Galois' and all the Demantes'. How many times had he tried to explain that the life he lived was *his* normal life, that he could not bear the abnormality of the life that *they* called normal? They would never understand. Why couldn't his sister speak simply and directly? Why must she pronounce her words like a bad actress on a provincial stage?

"Oh, Evariste! Some day you will understand. But then I fear it will be too late. Don't you see that by your actions you are shortening Mother's life? Stay with us when you come out of here. You will be welcome in our midst. With time, you and we will forget the nightmare of the present day."

To his sister, the prison (she never used the word "prison") was a disgrace and a terrible accident, a side-step from the broad and normal road of life into a deep and bottomless abyss. It was her duty to guide him back to the normal path through the shining light of family atmosphere.

Yes. The family atmosphere. How could he explain to his sister how he felt about the family atmosphere?

He said vaguely, "I must live my own life."

When he said these words, he was aware that he had absorbed his sister's style. His words sounded just as theatrical as hers.

His brother Alfred didn't say a word during this conversation. Evariste blamed himself now because he had never tried to discover what his brother thought and felt. Since his father's death he had remained aloof from his family. How obvious, how utterly transparent they were in their childish and insistent attempts to dominate his life. But today he had discovered something new about his family. He must have influenced his brother without knowing it!

Alfred looked at him with wide eyes. For the first time he saw how much adoration and love these eyes expressed.

He interrupted his sister's flow of eloquence to ask his brother, "Are you also ashamed of me?"

But again it was his sister who answered, "Oh, Evariste! Why do you say that? As though any of us were ashamed of you. Don't you understand that there is only one emotion in us, that of pity, and but one desire—to help you?"

Alfred waited until his sister had finished and then said quickly, "I am proud that you are my brother." Then, looking at the floor, he added, "When you leave Sainte-Pélagie I want to see you often and I want to talk to you."

If it were not for the iron grating between them, he would have embraced his brother. His sister bit her lips. She soon went away, asking him to think over what she had said.

"What is she doing now? She is telling Mother that they ought not to give up hope, that some day I may mature and change, but that in the meantime Alfred ought to visit me as seldom as possible, that Alfred is sensitive and will be susceptible to the bad influence of his older brother."

But Galois erred. His sister did not talk to anyone about Evariste. She confided her thoughts to her diary:

. . . No one could have imagined that the long preventive detention would not be enough to punish a slight misdemeanor. Again to pass months without breathing the air! What a sad outlook. And I feel very much that his health will suffer severely. He is already so tired! He abandons himself entirely to distressing thoughts. He is gloomy and old before his time. His eyes are hollow as though he were fifty years old.

Auguste Chevalier often came to visit his friend. Even his presence brought little relief. Galois resented any pity brought inside the prison walls by those who came from the outside. He saw through all attempts to conceal this feeling and responded by outbursts of anger or irony. With Chevalier it was worse. Here pity was strengthened by friendship and covered by a layer of mysticism and confusion, which grew around Chevalier as it did around the entire Saint-Simonist movement.

Auguste said to Evariste, "I talked about you with our Father."

Evariste interrupted: "You mean with M. Enfantin?"

"You know that to us he is our Father."

Galois murmured something which sounded like an apology and Chevalier continued, "I also talked about you with my brother."

"This time you mean your real brother?"

Chevalier answered meekly, "No, I really mean my brother in the Saint-Simon family. The fact that he is also my brother by blood is of little importance."

Galois asked with badly concealed irony, "Why did you bother them with my story?"

"When you leave this place, we would like you to spend a few weeks with our family. We invite you most cordially. You don't need to join us either now or later. But I'm sure it will do you good to come to Ménilmontant."

"You, Auguste, are the best friend I ever had or ever will have. But I cannot accept."

He thought how to phrase his reasons so as not to offend his friend.

"I am not made for family atmosphere. I am not made for discussions. No, Auguste. Thank you very much, but it would not work."

Auguste tried to argue.

"You say you don't like family atmosphere. But don't you see, Evariste, that this is a different kind of family? It is not held together by the mere accident of birth, but by the deepest kinship of hearts. We are bound by the same ideas, by the same beliefs and emotions."

"You say that it is a family by choice, by *your* choice. But if I go there I would have to accept it exactly as I have to accept my own family. No, I don't belong there."

"What you say seems logical and cold. You are always very careful to show that you are governed only by logic, not by emotion. But everyone can see through you very easily. You are governed by emotion more than any of us. Your attempts to show the sharp blade of logic do not deceive anyone. And it is this emotional life that brings you nearer to us, to the family of Saint-Simonists, than you would like to think."

Galois tried to answer calmly but became excited by his own words:

"Yes, you are right, Auguste. I am governed by emotion. But not the emotion you and your family like or cultivate. Mine is the emotion of hatred. Yes, my dear Auguste, I hate, I loathe, I despise. Yes, I know that you will tell me that I am capable of love too. I know it. There is no love without hate. Who has no power to hate has no power to love."

He looked at Auguste's suffering face and said, "I may be wrong, or you may be the very rare exception. But the idea of pure love is hateful to me. It leads to the muddled mysticism so apparent in all the writings of your family."

A few prisoners stood on one side of the iron grating talking to those on the other side, saying that they needed coffee, that they would like to see the children, or asking for warm underwear. In the midst of these small talks about the most important problems of their daily lives, Galois was delivering his oration against Saint-Simonism.

"The world is on fire and the Saint-Simonists discuss marriage versus adultery. But of course they don't say so. They clothe this trivial problem in mystical language and talk about love, family, priesthood, religion, motherhood, using noble-sounding words. But if you try to put what they say into simple language, the problem is merely whether a man ought to sleep with one or many women."

Auguste blushed but did not raise his evangelic voice.

"You are unjust to us, Evariste, and you know it. The problem is to destroy the family of blood and build a family of common ideas of love and justice. The family by blood can be destroyed only if we cease to know who our children are. It is not a problem of monogamy versus polygamy, but it is the most important problem of a family connected by blood relations versus a family bound by ties of common ideas and love."

Galois replied, "You live in an unreal world. You are isolated from the world which you neither know nor understand and which you believe will some day accept your leadership. The Pope, Louis Philippe, the Russian Czar will all bow their heads before your Father who will be the leader of the new world."

He felt that he had said too much. How far must he go to offend his friend?

Auguste said, "We do not deserve this irony. You may fight us as much as you wish, but we do not deserve to be treated like fools or lunatics. What you have said could have been said, and certainly was said, by many about Christ when he was alive. I know that we appear to you and to many others as dreamers. But even if we are dreamers, we have done a great deal of good for the people by awakening the conscience of the world to their misery. And I believe that ours is the future."

Galois felt worn out. He wanted to end the conversation, to lie down somewhere even if the bed was infested with bugs. He said feebly, "I'm sorry, Auguste. You don't know in what state I am, here in this filthy place. I'm sorry. I think I have lost my sense of proportion."

Auguste's eyes were moist. He spoke tensely:

"Soon you will be free. Come to us. Try it. Don't be too proud. Accept our invitation. Come to us. You will see that you will feel better. It will help you. I am sure it will help you."

Galois gripped the iron grating with his fingers until he felt a pain which seemed to bring relief to his burning head. His eyes were now two inflamed dark hollows.

He whispered, "Help! Help! No one can help. Only death."

VIII

FREEDOM REGAINED

1 : March, 1832

"ON APRIL 29 I shall be free! The end of my six-month sentence!"

By repetition of these words Galois tried to evoke longing and destroy apathy. He closed his eyes, hoping to see the glowing colors of Paris in spring, the *quais* on the Seine and the flowers at Bourg-la-Reine. But the pictures were two-dimensional and gray.

"Freedom regained! Freedom? There is no freedom in France—only tyranny. Paris—all France is one great Sainte-Pélagie. But at least I will fight and not rot in idleness."

He tried to think about the people, about the Society of the Friends of the People, about the Republic of France one and indivisible, about liberty, fraternity, equality, or death. But his thoughts were idle, apathetic, a lusterless repetition of old vivid thoughts now covered by a veil of dullness.

He wanted to evoke emotion, any emotion. He thought about Louis Philippe, waiting for hatred to possess him. But instead only phrases of vulgar abuse raced through his mind—words and sentences he had heard repeated hundreds of times in the courtyard of Sainte-Pélagie until they had become played out and meaningless.

Everyone and everything seemed annoying. His fellow prisoners, the Republicans? They smelled of brandy, they quarreled, they mixed patriotism and sex in filthy phrases, they made fun of his virginity, used words he had never heard before and explained their meaning by sickening gestures. If he could only run away and not listen to them! But their words and gestures evoked pictures that held him spellbound. He listened with burning cheeks and ended by hating himself more than he despised his fellow prisoners who had succeeded in confusing his mathematical thoughts by interpolating naked women between algebraic symbols.

"Of course not all of them are bad. Raspail! He is a great scientist and a great Republican."

No, he couldn't fully admire even Raspail. Why was he always writing letters, pretending that there was something to report from the stinking hole called Sainte-Pélagie?

"Only mathematics! How could I endure the dissipation and listlessness of Sainte-Pélagie if it were not for mathematics? There at least I am making progress. But the more I know, the more I discover, the more immense seems the unknown and unexplored terrain which lies before me. But that is always so. Newton expressed the same thoughts much better than I do."

He had spent eight months in prison for wearing the uniform of the dissolved Artillery Guard. During that time he had been abused by the authorities more than anyone else. A bullet had nearly killed him. Because it did not, he had had to spend one day and two nights in the dungeon. In January he was transferred to La Force for a week and then sent back to Sainte-Pélagie. Why was he sent away and why was he sent back? Why had abuse suddenly changed to pity? When Evariste, shivering with cold, entered La Force, even the clerk looked at him sympathetically, while listing the prisoner's dress: "Hat, tie, frockcoat, waistcoat, black trousers, wooden shoes; all half worn."

When he came back to Sainte-Pélagie the guards no longer bullied him and even the superintendent became friendly. Galois was too tired, too depressed and apathetic to wonder about the sudden change and its hidden cause.

Evariste was called to the office of the prison doctor, who listened to his heart, tapped his chest, and scribbled something on a piece of paper. The next day he saw the superintendent, and this high official was both understanding and fatherly. He cared, he said, about the well-being and happiness of all prisoners, but especially about that of Galois, whom he would like to help. He was very grateful, he said, that Galois' charming sister had drawn his attention to her brother's poor health. Yes, she was right, for it was shown on the doctor's report. He looked at Evariste with his small, blinking eyes.

"So, M. Galois, we have decided to do something for your good."

He stuck out his tongue and wet his lips while his hands rested peacefully on his extended stomach.

"You will spend the rest of your prison sentence in M. Faultrier's Nursing Home at rue de l'Oursine No. 86." He wet his lips again. "You will be very comfortable there, and the new atmosphere will make a new man of you. You will be allowed to do whatever you like, but you must not leave the nursing home until your sentence is ended. To this you must give me your word of honor." He closed his eyes and added with a dreamy smile, "All the arrangements are already made, M. Galois. You will leave Sainte-Pélagie tomorrow."

Galois looked more with disgust than hatred at the superintendent's sweet, sticky smile. Evariste asked himself, "Has he been bribed? Or am I so sick that he is afraid of a scandal if I die here?" He was too exhausted to answer these questions. He was relieved that he wouldn't see Sainte-Pélagie any more but felt too listless either for great joy or deep sorrow.

On March 16, Galois was transferred to the nursing home at rue de l'Oursine, not far from Sainte-Pélagie. He was conducted into a small room, where he met Antoine Farère, the young man with whom he would share it. After the cells at Sainte-Pélagie, the room with its two beds and a table between them seemed gay, clean, and full of light. His roommate looked very different from the political prisoners at Sainte-Pélagie. His blue frockcoat was cut by an excellent tailor and worn with careless elegance. The new neighbor greeted Evariste with a friendly smile, only slightly ironical, and Galois liked his long, handsome face, mostly for its softness, so different from the strong, hard faces at Sainte-Pélagie. His smile was fascinating, and his elegance was not annoying. True, Galois had seen many of his type, though perhaps few as good-looking and charming as Antoine. He had seen them in theater boxes, on horseback, or in carriages, in the company of pretty women—laughing, uttering their supposedly witty remarks, revealing excellent manners and good breeding, which for them was the essence of life and the road to success.

Galois had hated them all. But now, when for the first time he met one of their number, he found with astonishment that he was not repelled by Antoine. On the contrary, here at the nursing home he felt less apathetic, more alive and at peace than he had at Sainte-Pélagie. Evariste liked Antoine, who was neither boastful nor arrogant; who was cynical but intelligent, reserved but friendly, never quarrelsome, and always trying to conceal the superiority of his manners.

When Galois unpacked his belongings and put his many manuscripts on the table, Antoine collected all the little things that were spread over it and said, "You seem to need the table. It is yours."

"Thank you. Whenever you want to write I'll remove my things."

"Don't bother. I write very seldom. My friends have forgotten me and to my family I am a black sheep; they don't write to me."

He turned on his ingratiating smile and said to Evariste, "You seem to be a writer."

"I am not a writer, I am a mathematician." Then he added, "I was in Sainte-Pélagie for eight months."

"I was in La Force for a few weeks. Each of us seems to have a shady past."

Then he began to chatter with the charming irresponsibility of someone who knows that he is talking too much, but does so only to amuse and please his listener.

"You are a mathematician." He whistled. "Fancy that! A real mathematician." He whistled again. "Never saw one in my life before. Never knew that nowadays they put mathematicians in prison. I thought just the contrary—that one can be put in prison only for the lack of, but not for an abundance of, mathematical knowledge. This is exactly my case. When I added my bills they came to zero, but the man who put me here claims that they add up to eight thousand francs. I am here because I am a bad mathematician. It seems that one must be neither too good nor too bad. Each of us represents an extreme and therefore each is dangerous to the world. We must try to become average mathematicians, pool our knowledge, and arrive at a happy medium. This will be our salvation. It must be providence that brought us together."

He went on chatting about nothing, knowing exactly how much to say to be amusing and when to stop so as not to be a bore.

Galois found Antoine's superficial but friendly attitude relaxing. His cynical and empty talks contrasted pleasantly with the oppressive atmosphere at Sainte-Pélagie, where everyone wished either to destroy or to save the world in which he lived. When some days later Evariste confessed his Republicanism to Antoine in one tense sentence, his roommate was provoked to a long discourse:

"I am what you Republicans would call a parasite. I care little whether we have Charles X, Henry V—or whatever the name of the young brat is—or Louis Philippe or a Republic. No, that is not quite true. I wouldn't like a Republic, for it would take itself too seriously, it would not leave me in peace. They would start to talk about fraternity, equality, and virtue, virtue, virtue: Republican virtue. Phew! I would hate to be a virtuous man. I prefer a coquette any time to a virtuous Republican. I hope, my dear friend, you won't take this remark too personally. A virtuous Republican would not make love to the wife of his Republican friend. What a sad philosophy! France has perfected the art of love-making into a shining example for all Europe. This tradition must be preserved by all means."

Evariste anticipated a lecture on the art of love-making. But Antoine changed the subject of his monologue.

"Of course love-making is not a full-time occupation. It is not a sufficient remedy for ennui, the terrible disease of our age. For that we need cards or roulette in dim rooms. But to your friends you must pay card debts quickly, otherwise you are not a man of honor. This was exactly my predicament. I prepared a beautiful speech for my rich aunt. I nearly cried when I delivered my sermon—it was so touching. But, foolishly, I did not know that a pious priest had become friendly with her and convinced her that she must save me from hell below by letting me go through hell now."

"Then you did not pay your friends?"

"Be assured, my dear Galois, that you are sharing this magnificent apartment with a man of honor. I do not owe money to my friends; I owe it to a grasping moneylender to whom I signed many IOU's, renewed and renewed again, increasing with a terrific speed. You see I am a bad mathematician, but I am a man of honor."

Galois asked mechanically, "Was it worth while?"

He regretted the moralist's question and was afraid that Antoine would find it tactless. But the charming debtor went on with perfect ease:

"It was not as bad as you would think. For a short time I was at La Force, where I met a great number of interesting fellows. Then I became sick and was sent to a hospital, then here, where I have met an extremely interesting combination of mathematician and Republican. And now there are some indications that my aunt's heart will soften and her moneybag open. Then, with great regret, I shall have to leave you."

"What will you do then?"

"I hate resolutions and plans. In the worst event I can marry, have a family, and appear respectable."

Galois hesitated, then said, "I had thought before I met you that men like you ought to be hanged from the first lamppost, that bullets are too good for them. I still believe in the principle, but I would not like to see the prescription applied to you."

"The trouble with you Republicans is that you believe life is a terrifically serious business. No, it is not. I thought that Republicans are bores, troublemakers, bloodthirsty, that they ought to be shot. Mind you—not hanged, but shot—because they have lofty ideals. I still believe in the principle, but I would not like to see the prescription applied to you."

They both laughed.

Antoine had two visitors, both girls. Evariste was glad to be alone, to walk through the garden and watch the trio while concealing his own curiosity. He hoped and he feared that he might be asked to join them, and pondered gravely how to act and what to say. Then he sat on the bench, pretending to read, but his eyes wandered through the garden so as to sweep incidentally across the faces of the two girls. He knew the blond one, he had seen her before and had heard her laugh. But the other's face puzzled and fascinated him more. She had radiant black eyes which jumped restlessly from place to place as though trying to absorb and melt the objects on which they rested. Her black hair, parted in the middle, fell in ringlets against each cheek. It seemed blue when it reflected the sun. The beauty of her face was lessened by her somewhat too-thick lips and a slightly too-large mouth. With her lips half open to show sharp, even, white teeth, her face expressed primitive hunger as though she considered the world and all its pleasures to be her prey.

The three talked vividly, walking around the garden, then stopping, gesticulating, and resuming their walk. Evariste stared at the black-haired girl with an intensity that he no longer tried to conceal. What fascinated him most was the way her ringlets swung as she turned her head; the way she played with her little lemon-yellow parasol, running her long fingers over the slender black handle; how she slightly lifted her apple-green foulard gown; how the hardly visible motion of her hips formed tiny waves that sped down her skirt toward the green velvet band. All these wonders appeared to Evariste as the embodiment of grace and elegance, and not as the obvious, instinctively absorbed art of arousing desire. Suddenly the black eyes met his. The face seemed to light with a smile which was friendly, attractive, both promising and threatening. Then her eyes turned. But in that fraction of a second they set Galois' mind and body on fire.

When Antoine and Evariste were back in their room, Galois waited impatiently for Antoine to tell him about women, love, about the black-eyed girl, to hear a few words around which his imagination could spin a tale of romance. There was a long silence before Antoine spoke.

"Cholera morbus has broken out in London." He looked at Galois with half-closed eyes. "It has only to cross the channel to invade France. Paris is preparing a big reception. Cholera morbus is the talk of the town."

"Who told you that?"

Galois knew that it was a stupid question, much too obvious.

Antoine answered, "My visitors."

He laughed, and his eyes blinked ironically. Galois waited, but Antoine went back to the cholera.

"No one seems to know how it spreads. Is it infectious or not? Great doctors are of different opinions. You are a scientist, you ought to have your own view. What do you think?"

Galois said abruptly, "I don't know anything about it."

He waited for Antoine to begin again, but his roommate gazed dreamily and silently at the ceiling. The more Galois thought how to reopen the conversation, the more clumsy were his projects. He tried to appear casual, but there was too much hesitance in his words:

"You had two pretty visitors today."

Antoine looked at Galois and prolonged the suspense by letting him wait for the answer:

"No, my dear friend, I really had only one visitor."

"I distinctly saw that you had two." He pretended to be at ease, clumsily imitating Antoine's chatty style, "One of them was a blonde. The other had black hair and black, searching eyes. Even if I am a mathematician, I can distinguish one pretty girl from another."

"I still say that I had only one visitor. Jeanne, the blond one, is my friend. Yes, she came to see me today, she came before and probably will keep on visiting me as long as I am here, which I hope will not be for long."

He looked at Galois who hung on his words, smiled and added, "But the other one, as you say, the one with black, searching eyes, was more your visitor than mine, though you may not be aware of it."

Evariste rose, went to the table, turned his back toward Antoine, and pretended to look at papers covered with mathematical symbols. He still had not decided how to react when he half turned toward Antoine and said, "You are making fun of me."

"My dear friend, of course I would not make fun of you, for the simple reason that two men sharing a room ought to get along with each other as well as possible. And making fun would be of little help. Besides, it is not amusing. The pleasure of teasing requires both a victim and spectators. There is no sense in teasing if there is no audience. I hope, my dear Galois, that you are convinced."

Evariste responded gratefully to the cue:

"If you were not making fun of me, then please explain what you meant by your remark."

"The explanation is extremely simple. Jeanne has a roommate. Her first name is Eve; her full name, if I am not mistaken, is Eve Sorel. I have never seen her before and I don't know anything about her. Some days ago I told Jeanne something about you. She must have repeated your name and the very flattering remarks about my companion to her friend, who, unlike Jeanne, seems to be a Republican —a doubtful virtue, which I dislike, especially in women. I hope that I shall not offend you if I say that I could not touch a woman who is a Republican. I would be afraid that in the most intimate moments she might talk about royalists, the guillotine, the people's rights, and other similar nonsense. It is of course possible that even a Republican girl could forget the guillotine at some moments. But the fear that she might not would be quite sufficient to render me impotent."

He interrupted his own laughter when Galois asked, "I don't see how all this explains your previous remark."

"Excuse me, I went off the subject. Of course, I owe you an explanation. Eve seems to know a great deal about you and is very much interested in M. Evariste Galois. She told me at length about a trial at which she watched you. I did not know, my dear friend, that I had the honor of sharing a room with a man of your distinction. I am thrilled beyond description. A man who drank a toast to Louis Philippe with a dagger in his hand! Beautiful! Wonderful! Very courageous! Briefly speaking, Eve came here to see her hero. As I said, therefore, she was *your* visitor."

Antoine looked at Galois, who still stood half turned toward him and with unsteady fingers played with the pages of his manuscript.

"Eve is extremely anxious to meet you. She wants to see her hero face to face. I promised to use my influence with you. But, of course, if the idea is especially odious to you, please don't. And because of my high moral standards I am compelled to warn you that I don't know anything about her. Never blame me if you get into trouble, or if you find out that her Republicanism is not of the same brand as yours."

Galois knew how clumsy he was when he tried to hide his own thoughts and emotions. He could not hope to beat Antoine in this game of words.

"I would like very much to meet her," he said simply.

In August, 1817, the cholera had begun its march from the delta of the Ganges toward Europe. Fifteen years later it reached the gay streets of Paris. During those fifteen years it spread from its source toward Peking and the frontiers of Siberia. From there it traversed the snowy plains, crossed the Urals, and entered Moscow and Petersburg. It went with the Russian soldiers to their battlefields in Poland, more devastating, more feared than bullets and cannon. It did not distinguish between the Russian and Polish uniforms. It overran Poland, Hungary, Austria, and the Baltic ports of Germany. It jumped over huge areas leaving them intact, only to shatter the hopes of their inhabitants later by retracing its steps. In February, 1832, the cholera crossed from the ports of Northern Germany into England.

In Paris the sky was clear and blue, spring came early, and a dry refreshing wind blew from the northeast. The Parisians laughed, and some men went to the masquerades dressed defiantly as cholera morbus itself. One of them who danced and drank gaily felt a sudden chill and took off his mask. His blue face was more terrifying than the skull that had covered it before. He fell to the floor.

Those who did not run away but looked curiously upon his changing face later described the spectacle:

"His skin was blue, and you could have counted the muscles beneath it. His eyes were hollow, dark, and shrunk to half their natural size; he seemed like a corpse even before life had departed. His eyes sank in their sockets as if drawn by a thread toward the back of his skull; his breath was cold, his mouth white and humid, his pulse feeble, hardly noticeable, and his voice was a whisper."

On March 29, 1832, in Paris there was only one subject of conversation and one greeting, "Cholera morbus is in Paris."

2 : April, 1832

"On the twenty-ninth I shall be free."

Evariste stared at the grass, then slowly he dared to shift his eyes to Eve's shapely ankles. One of her legs swung in a slow rhythm, exposing and hiding the lace of her pantalettes. In desperate courage he raised his eyes to glance at the square-cut bodice where a puff of shell-pink tulle both concealed and revealed the edge of a valley between two softly rising mounds. He felt ashamed of himself for committing this sacrilege. When his eyes reached hers, they rested there, burning with humble devotion and begging for solace.

He knew that he should say something, but no thought came to his mind. The rising desire to confess his secrets and emotions choked away the words he wanted to say.

He repeated desperately, "On the twenty-ninth I shall be free."

They both sat on a bench in the garden of the nursing home. She looked at him serenely, half-smiling, her eyes offering assurance that whatever he might say would sound wonderful.

"It seems strange that soon I shall be able to walk along the *quais,* go to the Luxembourg Garden or to the Place Vendôme and to the glorious Faubourg-Saint-Antoine. I shall be able to go wherever I wish, and I shall see Paris again."

What Galois had just said sounded very stupid to him. He could have mentioned many other streets and places in Paris, thus prolonging the sentence indefinitely. The pause was long, and he felt relieved when he heard Eve's voice.

"You will not recognize Paris. The city is in mourning. People are dying of cholera by the thousands."

He did not want to talk about cholera, but it would be heartless to talk about himself. After hitting upon the subject of cholera, Eve did not wish to leave it.

"Some say that there is no cholera in Paris, that the government and the royalists poison wells, food, wine, and that people die of poisoning."

If any man had made this remark, Evariste would not have cared to prolong the silly talk. But now he was enchanted and grateful for an opportunity to expand his views.

"No, much as I dislike it, I must admit that this is perhaps the only calamity for which I cannot blame the government."

"Would you believe, M. Galois, that there were not enough coffins and hearses in Paris? A few days ago they started to collect corpses in both coffins and sacks, on artillery wagons. I, myself, have seen one of these wagons jolting so badly that the cords broke; the coffins fell on the pavement and some of them spilled the bodies. They were all blue. It was terrible."

She wiped the corners of her eyes gracefully with a handkerchief. Evariste felt a desire to kneel down and put his head on her lap. He would inhale the odor of violets, weep, and she would stroke his head.

"Now they collect the coffins and sacks in big furniture wagons. They are painted black and go from house to house. A man died yesterday in the house where I live." She smiled and said, "I shouldn't talk about such sad things. You have enough sadness in your life, I know. I thought that if I saw you, M. Galois, I might succeed in making you a little happier."

Evariste looked at her and in a low voice choked with emotion whispered, "I don't remember that I ever felt as happy as I do now."

She opened her eyes still wider and her smile broadened.

"What you just said is very beautiful."

At these words a stream of courage poured into Evariste's heart. He told her about Sainte-Pélagie, about his loneliness, and about the cruelty of the regime of whose many victims he was one. Then he said what he had planned for days and nights, always doubting whether he would dare to say it.

"I want to ask you something. I want to ask you a favor. When I am free I want to see you; I want to see you often. May I?"

He waited tensely, afraid that he would hear an excuse, or, still worse, only a half-hearted consent. The reply came quickly, and with it a relief to his unbearable tension.

"Of course we must see each other often. We must celebrate your freedom together."

He was overflowing with emotion. He told Eve that until now he had been interested only in books, studies, and political events. But now he felt a desire to live his own life, to start it on the day of freedom. He wanted to say more, much more, but his courage left him, though Eve's eyes shone with sympathy and understanding. As the end of the visiting hours neared, he became afraid that he had said too much, that he had only burdened Eve with his tense confession. Perhaps she would change her mind; perhaps the prospect of seeing him would seem repulsive to her.

He thought that he was staking everything when he asked, using her name for the first time, "Eve, we are friends, aren't we?"

Her eyes blinked excitedly when she answered, "Of course we are, Evariste."

He entered his room dreamily, He neither saw nor heard Antoine, who lay on the bed reading a journal with occasional loud comments.

"Good news for you and your Republican friends. Casimir Perier is sick. Cholera! You ought to be happy."

Evariste did not react.

Antoine murmured to himself, "There are many idiots in Paris. I wonder how many are not idiots. They believe that cholera is an Orléanist invention. Listen to this:

" 'At the street corners, near the red-painted wine shops, groups of people assembled, argued, and searched men who looked suspicious. They were doomed if something out of the ordinary was found in their pockets. The mob fell upon them like a herd of wild beasts. At the rue Vaugirard two men were murdered because they had some white powder in their pockets. I saw one of them as he gasped for breath. The old women took the wooden shoes from their own feet and beat his head until he was dead. He was naked and smashed to a pulp. His ears, his nose, his lips were torn out of his face. A wild man put a rope around the legs of the corpse and dragged him along the streets crying repeatedly, *"Voilà la cholera morbus."* A lovely girl full of fury, with naked breasts, and hands covered with blood, stood on the street, and when the corpse passed her she kicked it and laughed. She asked me then for a few francs so that she could buy mourning, because her mother had been poisoned a few hours before.' "

He folded and put away the paper.

"This is a very good description. It ought to end with 'Long live the People of France' or *'Vive la charte'* or something like that."

He became bored with his own words and the lack of response.

Evariste went to the table and looked at a sheet of paper. It was one of the pages of his unfinished manuscript. He sat down, dipped a pen in ink, and drew an elaborate monogram, E. S., on the margin of the page. Then he repeated it. Then he wrote "Eve," then "Eva," then "Evar," and finally in big letters he wrote joyously "Evariste."

Then he smiled happily.

3 : May, 1832

"I like the shape of these wineglasses. This line," Evariste ran his finger along the edge of the glass, "is a parabola, and by rotation about the axis you obtain a paraboloid—that is the shape of this glass."

Eve laughed.

"Is that how it was made?"

"Yes! Then they poured fluid gold into the paraboloid and it changed into wine. I like the mirrors and I like the red plush and all the riches."

(He thought, "My allowance is two thousand francs a year. It was immoral of me to come here and to pretend that I am rich.")

He said, "I would hate to be alone here. It would depress me. But today I love it."

("Will she understand why I said that today I love it? She could help me by one remark or by one question.")

He drank the wine and spoke loudly:

"The leg of mutton was excellent. I ate three hundred dinners at Sainte-Pélagie, all of them awful."

She said very softly, "You ought to forget Sainte-Pélagie."

The waiter brought chocolate parfait and coffee.

"I can't forget Sainte-Pélagie. It is the curse of my life that I cannot forget anything. Whatever I have seen and whatever I have read, all my experiences, everything sticks in my mind. Therefore if I hate or love, I cannot cease to hate or love, because the persons and events are always vivid before me."

("Will she help me now? If she would only ask 'Did you ever love?' Don't you see, Eve, that I need the help and encouragement of your beautiful, understanding eyes?")

He saw with relief that the man and woman at the table near theirs were leaving the room.

Eve said, "I am very different. I forget everything very easily. This must mean that I cannot hate or love."

"No! I don't believe you. I am sure that no one can love as tenderly and deeply as you can."

("Now I ought to say something more. There are so many things I want to say.")

He looked at her, but Eve's face became suddenly hard.

"You seem to know everything about me."

("Why did she change? It is my fault. I don't know what to say or how to say it.")

Eve interrupted the silence:

"Have you seen your Republican friends?"

"I've seen some of them."

("I am ashamed to admit to Eve how little I care now about Republican work. But she should understand.")

"I shall see my friend Lebon tomorrow. He is now the chairman of the group to which I belong. I need rest, I feel very tired. I don't think that I shall do any work for the Society in the next two or three weeks. I shall have a great deal of free time."

("If I had the courage I would tell you, Eve, that I want to keep all my time free for you!")

The waiter brought a bill, Evariste took out two gold pieces from his pocket, and Eve asked indifferently, "Then what will you do all day?"

("I could tell her that I will think about her if she would ask differently, if I did not have to decide how much to tip the waiter.")

"I shall work on my mathematical problems."

"Mathematical problems?"

He saw a spark of interest in her eyes.

("I must tell her how important my work is. She will believe me.")

"Before I went to Sainte-Pélagie I wrote a mathematical paper which I sent to the Academy. M. Poisson, a member of the Academy, had to judge it. It was sent back to me; he said that he did not understand it. He would, if he were a great scientist. I have new, very important results which I haven't written down until now. They are all here." He pointed to his forehead. "But I must write them down. Perhaps I shall succeed in making these foolish Academicians see the importance of my work before I am old or before I die."

(Eve thought, "The poor boy is crazy. He ought to be pitied. What do they want from him? Now he thinks that he is a great scholar. The Academicians and professors are just fools compared to him. But if he is crazy, then he may be dangerous. Didn't he act crazily when he raised the dagger? Who knows what he will do next?")

"Is there anyone who understands you?" she asked.

"Not one. I know that it is hard to believe; but you, Eve, will believe me. There is not one human being who understands what I have done. There are in the whole world a few men who could understand my work, but they don't know it or don't want to know it. And there are one or two who believe in me although they don't understand my work."

("He is crazy, poor boy, and he suffers. I am sorry for him.")

Her eyes lit up with pity which Evariste took for trust and confidence.

"Who believes in you?"

"Very few know that I am a mathematician. I don't like to talk about it. But it is different with you."

("She looks at me with sympathy. I am too impatient. Her love may come some day.")

"I have a great friend, Auguste Chevalier, who believes in me. He is perhaps the only man who believes in me as much as I do."

"Who is Auguste Chevalier?"

"He is a wonderful man. He is a Saint-Simonist. Perhaps a little queer on the subject of saving the world, but otherwise the most noble and lofty man you can imagine."

("The only man who believes in him is queer by his own admission. He is crazy. But he has beautiful eyes, deep and burning.")

"And is he the only one?"

"My father did. He took his own life nearly three years ago. He believed in me."

("Was the father as crazy as the son? The poor boy almost has tears in his eyes.")

She asked sympathetically, "But you had teachers who knew you. Didn't they believe in you?"

"There was only one who did—M. Richard at Louis-le-Grand. When he found out that I was a Republican he tried to convince me that I ought to care only about mathematics. He thought I was crazy to believe in the Revolution and in the rights of the people. Since then I haven't seen him."

("Now he tells me that someone else thinks he is crazy. The poor boy trusts me. I can do whatever I like with him. It is too easy to be amusing.")

She smiled. Evariste was happy to see the friendly smile.

"Am I boring you with all this talk about mathematics? I wouldn't have said anything about it to anyone else." He hesitated, and added desperately, "Everyone else would think that I am crazy with conceit. But you believe me."

"Yes, Evariste, I do believe you."

Evariste went from his room on rue des Bernadins toward rue de l'École de Médecine, where his friend Nicolas Lebon lived. He still felt the pleasure of wandering at will through the streets. In one year Paris had become older, more reserved, depressed. Many women were in mourning, and occasionally coffins were still collected by furniture wagons. But the epidemic was subsiding and Paris had become bored by cholera, which ceased to be a subject of fashionable conversation.

Evariste turned toward rue des Noyers and walked slowly, looking hungrily at walls and faces. He stopped when he saw two outdated proclamations, side by side, which miraculously had survived on that spot for more than a week. Evariste read one signed cryptically, "Republicans":

> For two years the people have been a prey to agonies and the most disgraceful miseries; they have been attacked, imprisoned, murdered.

And this is not all; for under the pretext of a pretended epidemic, people are poisoned in the hospitals and assassinated in the prisons. What remedy is there for our ills? Not patience, for patience is at an end. No! It is by arms that the people can gain and maintain both their liberty and their bread.

He had had enough. He felt disgusted and ashamed, hoping that this stupid, provocative manifesto signed by Republicans had never been written by Republicans. He turned toward the other proclamation, signed by the police:

In order to convince people of their atrocious accusation, some wretches are visiting public wells, wine shops, butchers' stalls with pockets of poison to empty into the fountains and wine and on the meat; or they even pretend to do so. Then they cause themselves to be arrested by their accomplices who, after identifying them as attached to the police, stage the escape of the accused and so try to demonstrate the truth of the odious charge brought against the authorities.

Evariste read the proclamation twice in order to absorb its incredible perfidy and to whip up his own anger over the accusation and counteraccusation. But the fire of indignation and hatred was short-lived; it had lost the consuming power that it had had a year before. He thought about Eve.

The medical student greeted his friend with an enthusiastic torrent of words:

"I went twice to your place and couldn't find you. What are you doing now? You received my letter, of course. How have you been since I last saw you? Tell me everything. How was it at the nursing home? Do you feel well now?"

He spoke loudly, with vivacious gestures contrasting with his fat lazy frame.

"I am very happy to see you free again. In two days we shall have a meeting here. This is what I wanted to tell you. We are all very anxious to have you with us again."

He calmed down and listened to Evariste:

"As to my last weeks, there is not much to report. I did not see any Republicans in the nursing home, and in prison there is only depressing ennui, events which are important only to the inmates."

"No, I don't agree with you. Sainte-Pélagie is important to us because it is our fortress. Three weeks ago, as you know, a gang of spies and *provocateurs* attacked that fortress and killed one of our patriots. They want to get rid of us without dragging us into court. You see how clever they are. They killed a patriot and they claim that a Republican mob did it while attacking the prison. They are devilishly clever now. The bastard Gisquet must be hanged from the first lamppost. But tell me how you feel. In the last year you have changed very much. You are very thin. I hate to say so, but you look very tired and even older than I do. What happened to you?"

"I can't hide from your sharp medical eye that I feel worn out and tired."

"I had planned to ask you to start your work with us immediately. We have great hopes now. In June or July something usually happens in France. And this year it will come, if the weather is good. You can't have a revolution on a rainy day." He laughed loud. "There is a lot of inflammatory material. Much more than a year ago. All we need is one good lucifer match to light the powder and the explosion will jump sky-high." He threw up his hands, violently indicating the tremendous explosion.

"My dear Nicolas, I remember a year ago you said something very similar."

"Yes, I know, I am the official optimist. I may have said it a year ago, but this time I shall be right. We need one good spark and the explosion will come. We have done much work on educating the people."

"I'm glad you think so."

"My dear Evariste, many things have changed during the year while you were away. They became more ruthless and determined with every month. But our strength grew too. On the one hand, the cholera. Then perhaps the most important and fortunate thing: Casimir Perier is dying. He may be a corpse at any moment. Now, it will not be easy for the King to replace him. This damned son-of-a-bitch was strong, there is no doubt about it. He kept all this disintegrating rot together. What we need now is one issue, one event; as I said, the lucifer match, and the huge fire of a revolution will burst out by itself."

He looked at Evariste again and said, "But I am worried about you. You have changed. I had thought that you would be with us immediately, but now I wonder whether you ought not to rest for two or three weeks before you start anything."

"I'm glad you said that. I don't feel strong enough to start work. I came to tell you this but I'm glad you mentioned it first."

"You seem so damned refined about everything. You must have met some well-brought-up people lately. Anyhow take my advice, both as your friend and as a medico. You need a rest. You won't have it in Paris. Go to the country. When you come back you will be doubly useful. But go away immediately. We may need you very soon. Promise me that you will leave Paris."

"Perhaps a little later. I cannot leave just now."

Lebon looked at him in silence and then asked, "Woman?"

Evariste nodded and looked at the floor in embarrassment.

Lebon said, more to himself than to Evariste, "I am surprised though I don't know why. It is certainly the most natural thing in the world. But I still don't see why you are so damned refined about it."

It was almost a month since he had met Eve for the first time. He could now stroke her face and her smooth black hair as long as he did not disarrange it too much. He could touch and kiss her cheeks, mouth, neck, down to the sharp line of her dress and catch an exciting glimpse of her breasts. Once or twice he even fondly touched them through the silk dress. (But there was not even a trace of encouragement this time.) He stroked her legs below the knee to confirm what he knew, that one of them was as well shaped as the other. Maiden and unexplored land lay before him. Its vision and the hope of its possession through love and love alone kept him awake at night, kept him lazy and worn out during the day, filled with dreamy anticipation, plans of conquest, and fear of defeat.

("Does Eve know that I love her? Does she love me? How otherwise could she allow me to touch her hands and kiss them? Why then does she always silence me when I try to tell her about my love? The last time I collected all my courage and asked her, 'Do you know what I feel toward you?' why did she cover her mouth with her palm which I humbly kissed? I didn't dare repeat my question. No, this cannot last any longer. I must have clarity. Clarity!")

"It is a month today since we met. This is our celebration at the same place where we had our first dinner together."

Evariste raised his glass.

("In this one month I have spent four times my allowance. Fashionable restaurants, cafés, good tailors, hatters, all cost money. Did she notice my new jacket?")

Eve said, "There is another cause for celebration. It is the day of Perier's funeral."

"I don't like to celebrate death even if it is Perier's. His funeral was a sorry spectacle. No one was moved."

"What will happen now?"

Evariste looked obstinately at the table and without raising his eyes said, "I want to talk about us."

(She thought, "You sound crazy when you tell me what a great mathematician you are, you are a bore when you tell me what a virtuous man Robespierre was, but you are the most clumsy man in the world when you try to make love. I like your eyes and I am still sorry for you. But mostly I hate you for treating me like a saint and a virgin. Your own stupidity is your greatest enemy.")

"I want you to listen to me, Eve." ("I must stick to my resolution.

Last night I couldn't sleep. I decided to tell her and I must do it. I must have courage. I shall speak whether she helps me or not.") "Will you listen to me, Eve?"

("He has made up his mind and there is nothing I can do. Like an obedient marionette, he moves when the strings are pulled. I must take orders from the other bastard. But I could have said that the plan doesn't work. It is too disgustingly easy. Why is he so stupid—the poor boy?")

"Of course I will listen to you, Evariste."

He raised his eyes from the table, looked gratefully into hers and said, "Thank you, Eve."

("It is too late now to withdraw. Now I shall have to tell her.")

"I couldn't sleep last night. All my life I have striven for clarity. I can think for days and nights about my mathematical problems, trying to see their solutions clearly—when I speak to my friends, when I eat, when I listen to the speeches of my Republican friends. Even when I sleep, my mind works for me; and sometimes I wake up and suddenly I have before my eyes the solution for which I have been searching for weeks. I have always looked for clarity." He paused, absently drawing ellipses on the table with his middle finger. "I have worked very little in the past month. I also withdrew from Republican work. When I do not see you, I waste the hours of my days in dreamy meditations. This can't go on. I can't bear it any longer."

("He is so moved that he can hardly talk. He is waiting for help. No, I am not sorry for him. He is stupid.")

"I am very much distressed to hear it, Evariste. I never dreamed of taking you away from your important work—whether Republican or mathematical. When you first told me about your passion for mathematics, I said to myself 'I am happy to help a great scholar find some relaxation. He will work harder and better.' "

("Does she not understand what I want to tell her? Perhaps what I am going to tell her is not true.")

"Eve, you don't understand! You seem to think that I blame you.

What I mean is that always in my life I have striven for clarity, as I told you. But the relationship between us is the antithesis of clarity. And this disturbs my days and nights until the happiness of our first hours has changed into unhappy brooding and melancholy. I must have clarity. I must know how you feel. I love you, Eve."

("Is it true that I love her? I never thought about my love without thinking about hers. I imagined all possible answers. But there was always love or at least hope of love in her answer. It is different now. I feel it, I am sure of it. No, she doesn't love me. But why? Something went wrong. When and where? Why? Everything always goes wrong in my life.")

He looked at the pictures, mirrors, chairs; their shapes became fantastic. And Eve was saying, "We have known each other only one short month. True, we saw each other often and we had a pleasant time together and I always liked you. But still you must admit that we know very little about each other."

She wanted to say something more, but Evariste interrupted excitedly:

"I know what you want to tell me. That we shall be friends, but that you will never love me." There was belligerency and abuse in his words:

"You will love me as a brother. There is hardly a cheap romance in which the heroine does not offer her sisterly love at some time or another. It is very kind of you to sugar-coat the humiliating truth. But I don't need pity."

("He is capable of anything. I shall not allow him to make a scene here. Not here.")

"No, Evariste. That is not what I wanted to tell you."

"What then? What is it? Please, Eve, tell me. Perhaps I ought not to have said what I did. But if you knew what I suffered, you would forgive me. Please, Eve, tell me."

"I wanted to tell you that I never thought about you in the way you think about me. What you said was sudden and unexpected. I really don't know what to say. I will have to think over everything you've said. Perhaps tomorrow I can tell you more."

("There is still hope. Perhaps I acted too hastily. It is only one month since we met. Perhaps there is a spark of love in Eve. Perhaps she doesn't know it herself but it may grow. Perhaps there is someone else. Why didn't I think of it before? There may be another man.")

"Is there someone else perhaps? If so, please tell me, Eve."

Now hers was a suffering face when she said, "Please, Evariste, don't ask me anything. Tomorrow afternoon I shall come to see you and then I will tell you everything. Let's not talk about it any more today. Please promise me."

The little hope he had had was drowned in despair. He said apathetically, "As you wish, Eve. I promise you."

Evariste went to his room, lit a candle, threw himself exhausted upon the torn red-plush armchair. He had done what he had decided to do; he had told Eve that he loved her. Never before had he felt as little love as now. He felt more hate than love. And yet he knew that love would come back in a mighty torrent if he heard one tender word from Eve.

"If she humiliates me, I'll hate her, I know. I can't bear my love without hers. I don't want only to love. I want to be loved too. What will she tell me tomorrow? Perhaps she will tell me that I must be patient, maybe in a year or two she will see. Or perhaps she will tell me that there is someone else—some empty, stupid man who has money and knows how to dress."

For the first time, the Eve he saw was different from the Eve he had created.

He undressed mechanically.

"Why can't I go through life without a woman? Mathematics! There is purity and beauty in mathematics. It will never disappoint me. Perhaps I know nothing about Eve. Why do I always make a mess of everything? Why does everything go wrong in my life? Self-pity again. How I hate it!"

He put out the candle and lay on his bed.

"There is still hope. Tomorrow is not far off. We shall see. If not, there is mathematics and there is the people's fight. What does it matter whether Eve loves me or not? Why did I think that Eve understands me, that she is different from others? Is it not because I wanted her to be different? Perhaps no woman could ever understand me. I must learn to be lonely. No woman—to be alone—like Newton. Great men have been lonely. Not by inclination, but life taught them to be lonely. And life is teaching me, too. I must accept this lesson with humility. But two things remain for me: the people's fight; mathematics."

When Eve entered Evariste's room she stood at the door rigidly; her eyes were hard. Galois pushed toward her the only armchair in his room; its worn-out red plush exposed white spots of stuffing.

Evariste looked at her closed mouth and stiff back and wavered between a desire to throw words of abuse at her and to beg for a tender embrace with words of love.

He said, "How did you sleep?"

"Very well, thank you."

("I have never before seen this stony, cold face. What does it say? Not love, not indifference, not even hatred. I don't know. I know only that it will be different from all the possibilities I have imagined. Two days ago I touched her hands, stroked her hair, and kissed her mouth. I know that I will never do it again. I have no desire to do it again. But I do desire to throw her on my bed, to use force, to see fright in her eyes, to humiliate her, if she humiliates me.")

"I promised you yesterday that I would come here. I came to tell you that we see each other for the last time."

Evariste became frightened. Mechanically now, he spoke humbly, amazed how his words contradicted his mood of only a moment before:

"But Eve, why? Something has happened between us that I don't understand. I must have done something wrong. Perhaps I was too hasty and I should not have said what I did yesterday. I don't understand. Why are you so changed? If I did something, if it is my fault, please tell me. Perhaps I can change it."

He finished abruptly, not knowing what else to say.

Eve sat calmly and then spoke very slowly, hardly opening her lips, her face hardening with every word:

"You told me you want clarity. All right, you'll have it. I am the mistress of a man I like very much. He is a patriot. My lover was away from Paris for six weeks. I was glad then to have someone take me to good restaurants and cafés and teach me the history of our revolutions. I didn't object to your kisses and might have given you more if you knew how to ask for it. On the whole I liked you, though your technique is clumsy even for a beginner. But after all you are a mathematician, and no one expects a mathematician to be a great lover. In a few days my friend is coming back to Paris, and I can't and won't want to see you again. I'm sorry I had to make things so clear to you, but you asked for clarity and I hope you're satisfied now."

"You're lying! You're lying! It's not true! It can't be true."

She looked at his face from which the blood was drained, with hard, impertinent eyes.

"Do I look *now* like a woman that's lying?"

He stood up. Eve sat opposite him, her tight-fitting pelisse still buttoned to her throat. Her arms rested nonchalantly on the arms of the chair. Evariste did not know for what reason he rose, whether he wanted to beat her, or strangle her, or use his dagger made by the cutler, Henry. He felt a mixture of all these desires. But they all suddenly collapsed. There was no fright in Eve's face. Her mouth, which he had kissed, now seemed big, greedy; the bright black eyes stony, implacable. Her face was ugly, spiteful, cruel, a symbol of sin and debauchery.

Evariste stood before her and cried, "Then you are a common slut, a prostitute anyone can have. You played with me as you have played with hundreds of other men. I was the only one stupid enough to think you innocent and capable of love. Yes, it's very funny. I was involved with a common infamous cocotte, a prostitute. Perhaps you take money, too. If so, tell me how much I owe you. Whores take money for what they do, don't they?"

He shouted words of abuse; vulgar, filthy words which he had heard at Sainte-Pélagie.

Eve rose. Two red spots glowed on her cheeks, and there were fury, scorn, and hatred in her face.

Her loud voice covered Evariste's outburst:

"You feel very superior, M. Galois, don't you? I am low, but you are the great and noble man, the innocent, the friend of the people. Now let me tell you something. You talk about things you don't understand and never will. I am uneducated, vicious, low; and you are educated, noble, a great mathematician, a son of a mayor who fed you chicken and white bread. You dare to speak to me the filthiest words I ever heard."

Evariste clenched his fists and cried, "I will kill you if you talk about my father."

"Kill you." Scornfully she imitated his voice. "You can't even kill. You learned about killing from books too. You can only talk and talk. I was never afraid of you and I shall never be. I'll tell you more. You'd better be afraid of me. Because I swear to you that you will regret the words you've said to me. Yes, M. Galois, you will regret them. These are my last words to you."

Evariste heard a sharp bang of the door and quickly receding steps. He was alone. He stared at the red armchair on which Eve had sat. He threw himself toward it, knelt on the floor, and tore the plush that covered it, throwing big pieces of cloth and the cotton stuffing all over the room. He tried to smash to pieces the wooden skeleton, but succeeded only in breaking off one leg, and threw the rest into the corner of the room. Exhausted, he threw himself down upon the bed.

Tears fell from his eyes; heavy tears which washed away but a small part of the burden of his life.

4 : May 25, 1832

Forty disciples of Saint-Simon followed Father Enfantin to Ménilmontant in April, 1832. Among them were the brothers Michel and Auguste Chevalier. Poets, musicians, artists, scientists repaired the house, swept the rooms and yards, cultivated the grounds, and covered the walks with gravel. When, at five o'clock in the afternoon, the horn announced dinner, the members of the Saint-Simon family piled up their tools, took their places at the tables, and greeted Father Enfantin with the words, "Hail, Father, hail. Salutation and glory to God."

There in Ménilmontant, Auguste Chevalier received a letter from Evariste with words embittered by suffering and confused by disappointment. He understood little of its contents and knew but one remedy: for Evariste to join the Saint-Simon family, to be among those who loved him. On May 25, Evariste again wrote to Auguste:

My dear Friend,

There is pleasure in being sad, if one can hope for consolation. One is happy to suffer if one has friends. Your letter full of apostolic grace has given me a little calm. But how can I remove the trace of such violent emotions as those which I have experienced? How can I console myself when I have exhausted in one month the greatest source of happiness a man can have? When I have exhausted it without happiness, without hope, when I am certain I have drained it for life?

Oh, how can you preach peace after that! How can you ask men who suffer to have pity? Pity, never! Hatred, that is all. He who does not feel deep hatred for the present day, cannot feel love for the future.

I approve of violence—if not with my mind then with my heart. I wish to avenge myself for all my sufferings.

Apart from that, I am on your side. But let us leave all this; there are perhaps those who are destined to do good but never to experience it. I believe I am one of them.

You tell me that those who love me want to help me and to smooth out the obstacles which life has put before me. You know how rare are those who love me. This means that you feel it your duty to bend all your efforts toward converting me. But it is my duty to tell you again, as I have done hundreds of times before, that your efforts are in vain.

I wish to doubt your cruel prophecy that I shall not do research any more. But I must admit that there may be some truth in it; to be a scholar one must be only a scholar. My heart rebels against my head. I do not add as you do, "It is a pity."

Forgive me, dear Auguste, if I have shocked your filial feelings by referring improperly to the man to whom you are devoted. My remarks were not spiteful and my laugh was not bitter. This is quite an admission on my part, considering my present state of irritation.

I shall see you on June 1. I hope that we shall see each other often during the first fortnight in June. I shall leave around the fifteenth for Dauphine.

Yours

E. Galois

P.S. In rereading your letter I notice a sentence in which you accuse me of being intoxicated by the putrifying filth of a rotten world which soils my heart, head, and hands.

There are no stronger reproaches in the vocabulary of men of violence.

Intoxication! I am disenchanted with everything, even love of glory. How can a world which I detest soil me? Think about it!

5 : Tuesday, May 29, 1832

On Monday Evariste returned home late at night. On the floor he saw two visiting cards and a letter, slipped into his room under the closed door. He lit a candle, took the two visiting cards, and looked at them for a long time. Each bore an identical message; only the names and handwritings were different:

Pécheux d'Herbinville
will be in M. Galois' room tomorrow
the twenty-ninth at nine in the morning.

Maurice Lauvergnat
will be in M. Galois' room tomorrow
the twenty-ninth at nine in the morning.

He turned the cards over a few times and put them into different relative positions. He saw an outline of a face appearing on these cards. He closed his eyes to erase the picture, but the face slipped in between his eyelids and his brain.

He opened the letter. Again a half-transparent face stood between his eyes and the paper; it made reading difficult. His friend Antoine wrote him in his cynical and amusing manner that his aunt had softened and paid his debts, that he was now free and would come next day late in the morning to see his companion from the nursing home.

It was seven o'clock in the morning when Evariste woke up. He dressed, and on his way down asked the janitor's wife to clean his room, since he expected visitors. He went to a near-by café which was almost empty. Apathetically he looked around while eating breakfast. He glanced at his watch, a good gold watch his father had left him. It was eight o'clock. He paid and walked back toward his home. On his way he noticed a proclamation which he had not seen before. It was signed with thirty-four names: Lafayette, Odilon Barrot, Laffitte, Charles Comte, and others. These men now appealed to the nation, acknowledging Louis Philippe, but asking him to change the course of his actions. Evariste decided that the proclamation was weak, dull, and watery beyond description. How many times had he heard all these sickening phrases about Belgium, martyred Poland, the foreign and internal policy which must be changed if the government wished to survive and be loved by the people of France.

"An obvious, too obvious, offer to Louis Philippe, now that Casimir Perier is dead and buried." He repeated in his thoughts, "Dead and buried."

He read the closing sentence: "The France of 1830, like that of 1779, believes that a hereditary monarchy surrounded by popular institutions is not inconsistent with the principles of liberty."

"An offer by men who want to be lackeys." But there was no bitterness in his thought; only disillusion and apathy.

He went back to his room. The bed was made up and the floor swept. His table was full of papers. He had asked the janitor's wife never to touch them. Now they lay in disorder. He piled them up, looked once more at the visiting cards, and then threw them upon the clear surface of the table. He went to the windows and saw, through the half-transparent curtains, two men standing motionless opposite his house. In one of them he recognized Pécheux d'Herbinville. The other, tall, carefully dressed, had a large, square face familiar to Evariste. He remembered that he had seen him at a public meeting of the Society of the Friends of the People and at the banquet at Vendanges de Bourgogne, both times sitting with Pécheux d'Herbinville.

"They are aristocrats who out of boredom became Republicans and are as proud of their manners, and even their ancestry, as the bourgeoisie of their possessions. Crossing the street will take them eight seconds, climbing the stairs twenty seconds, then they will wait before my door, and exactly at nine they will knock with their aristocratic fingers."

"Please come in."

They entered. Galois rose from his chair. They bowed stiffly and Pécheux d'Herbinville said, "My friend Maurice Lauvergnat and I have come to you in an affair of honor."

He spoke very precisely and slowly, a lesson learned by heart and rehearsed many times. He still curved his lower lip when he emphasized words, exactly as he did the day Galois saw him speaking before the City Hall and at the trial of the nineteen.

Galois bowed slightly and did not answer. He seemed ridiculous to himself when imitating the good manners of these aristocratic Republicans.

"During my absence from Paris you were seen often in the company of Mlle. Eve Sorel. She told me that she saw you on your insistence, out of pity and perhaps even out of sympathy. You misused her sympathy. Knowing the relationship between me and her, you tried to influence her to leave me by slandering my name, by telling atrocious lies about me. After you had discovered the futility of such methods, when your attempts to seduce her failed, you poured vulgar and indecent abuse upon my friend. M. Galois! I wish to tell you in my name and in the name of my friend M. Maurice Lauvergnat that you acted dishonorably. I came here yesterday and again today to challenge you to a duel." Nonchalantly he threw a piece of paper on the table and added, "Here are the names and addresses of my seconds. They will expect yours."

Then Maurice Lauvergnat said in a hoarse voice, "As a Republican, a patriot, as a friend of M. Pécheux d'Herbinville, and as a maternal cousin of Mlle. Eve Sorel, I challenge you to a duel which I am ready to fight with you at any time after the one with M. Pécheux d'Herbinville is terminated."

Galois answered calmly. His voice was almost as composed as those of his adversaries, his words almost as measured as theirs.

"Gentlemen: Upon my honor as a Republican and patriot I swear to you that you shall have the truth. I do this because I want to avoid a duel in which the death of at least one Republican seems certain. I do not wish to die, and still less do I wish to kill for an unworthy cause. As to your accusation, I admit I have been in love with Mlle. Eve Sorel. I did see her during this month. But I assure you, gentlemen, that I did not know anything about the relationship between her and M. Pécheux d'Herbinville until last night when I saw the visiting cards of both you gentlemen, and sensed a connection between Mlle. Eve Sorel and the affair of honor indicated to me by your cards. It is, however, true that I used abusive language to Mlle. Eve Sorel."

He felt how weak it sounded. What more could he say? If he blamed Eve for what he had said, he would be regarded as doubly dishonored. These two Republican aristocrats—one with the curved mouth and the other with the hoarse voice—these tailor's dummies regarded abusing women, and especially their women, as a crime much greater than the betrayal of their country.

Evariste decided to add only one sentence:

"This I regret, and I am ready to apologize."

Maurice Lauvergnat answered quickly, "There is only one way you can apologize for such behavior: by swords or pistols."

Galois replied with unshaken calm, "I want to avoid bloodshed. If you knew the full story, you would also know that I was provoked into using the language I did. I repeat that I am ready to apologize. What more can I say?"

He hesitated whether to say more. But Lauvergnat's hoarse voice cut in:

"You are a coward, you want to avoid the duel by dressing yourself in a Republican toga. But at the same time you act still more dishonorably by intimating that Mlle. Sorel provoked your abuse."

Pécheux d'Herbinville seemed slightly embarrassed at his friend's outburst of anger.

Evariste lost his calm, not by degrees but suddenly in an abrupt transition. His controlled contempt and bitterness burst into words which, like pistol shots, he threw at his adversaries.

"Coward! Coward! What an easy accusation. According to the code of honor, I am expected to react violently and make you the offended party. I am expected to be indignant over your accusation and prove by the evidence of my own dead body, or by the evidence of your dead body, that I am not a coward. I must demonstrate that I am not a coward to you gentlemen about whose judgment I care nothing. Otherwise you will proclaim to all Republicans, to all patriots, that I am a coward who refused a challenge.

"You will not tell my friends that all I did was to use violent and, let us say, abusive words in a fit of anger toward a woman of doubtful virtue, who cynically ruined my life. You compel me to die for a detestable cause. You want to murder me because a vicious girl lied to you. I take heaven to witness that I told you the truth. If you still insist, I am ready."

Pécheux d'Herbinville said icily, "What you have just said would be sufficient reason to challenge you if I had not done so already. Since I did, there is nothing for me to say. I await the arrangements of my seconds."

They both bowed and left the room.

Evariste went to the window, opened it, and looked out at the street. He saw a fat woman opposite his house, the shoemaker's wife, standing before her husband's workshop. A small, thin girl with smooth black hair stood beside her, and the fat woman tenderly stroked the girl's head.

"This fat, untidy woman loves her little girl. Perhaps tomorrow she will again stroke the head of her child, or perhaps she will scold her. I don't know what she will do tomorrow. I shall never know, I shall not be here to see. I shall be dead."

He saw that his two visitors were entering a carriage.

He looked at the people, walking in all directions, gesticulating, talking, arguing. At a stall opposite his window, he saw a woman pinching one cucumber after another.

"She is discussing now with the peddler the price and quality of cucumbers. They are alive! They will be alive tomorrow. In fifty years nearly all of them will be dead. But the earth, the buildings, the stones of the street, the whole external scenery may remain unchanged. New men will enact a new play on the same background. The sun will shine again, the green earth will be green again. The body of the woman who strokes her child's hair, the body of the woman who touches the cucumbers, the body of the old man who argues with her, they will all rot; they will all be dead. Their time will come later; my time will come tomorrow.

"My thoughts are melodramatic and stupid, like a bad drama. M. Hugo would do it much better."

He smiled feebly.

"Two men want to kill me because of a woman I have offended. There is no anger and no bitterness in my heart. I tried to tell them the truth. Of course it didn't work. How could it? Why did I deliver a sermon to these men? Because I like to make speeches. I have made my last speech.

"Where is the hatred that grew in my heart? The breath of death melted my hatred." He repeated to himself, "The breath of death melted my hatred. I am at peace. I have longed for and waited for peace. I am not yet twenty-one, but peace has come to me arm-in-arm with death."

He looked out the window. The colors of the world around him seemed more vivid than ever before. The people of the whole world smiled at Galois and saluted him.

Someone knocked at the door. It was difficult for Evariste to take his eyes from the street.

Antoine came in. He began gaily:

"So here we are; here you are, and here you live. Now we are both free. A patriot and a parasite, they meet again. You look calm and sad. What has happened to you?"

"I'm glad you came. You came just at the right time. Today I was challenged to a duel by two patriots—a duel of a purely personal nature. I would like to keep my Republican friends out of this miserable affair. I ask you to be my second."

Antoine's face was expressionless.

"Could you tell me a little more about it?"

"Two men, whose visiting cards lie on the table, challenged me in defense of Eve's honor."

Antoine gave a long whistle. Now he looked embarrassed as he said, "I seem to be responsible for all that. But, my dear friend, as I told you, I saw her only once in my life and I don't know anything about her. Judging by her roommate—you remember Jeanne, the blond one—there can't be much honor in her worth fighting for. Luckily I got rid of my girl without any duels—at least so far."

Evariste stared at the open window and threw his words into the street without looking at Antoine.

"If you could convince my adversaries that, as you say, Eve's honor is hardly worth my blood or theirs, I would be very happy to withdraw from this useless and silly duel."

He turned toward Antoine and spoke impatiently:

"Do what you can, everything possible, to reconcile me with my adversaries. I'm ready to apologize for what I said to Eve. My attitude may be shocking to you, but my ideas of honor are different, and I don't care if these gentlemen think me a coward. I want to avoid this fight. Do you understand?"

"I think I do, and I can promise you to do my best."

"Here are the names and addresses of the seconds."

Antoine turned over the piece of paper a few times.

"I am afraid that from what you tell me your adversaries do not mean a duel in which everything is carefully staged, even to the gestures and words of reconciliation."

"No, I'm sure they mean business."

Antoine whistled and then said, "I can assure you that I shall do my best to represent you properly and according to your wishes. Who is the offended party?"

"I don't know, because we offended each other. You will have to settle that with the seconds. I don't care about the details. My theoretical and practical knowledge of duels is nil."

"If the choice of weapons belongs to us, what would you choose? Swords or pistols?"

"Pistols."

"In that case you need two seconds. Let me relieve you of all such worries. I'll be glad to ask one of my friends to help me, and we'll arrange everything."

"Thank you. It is very kind of you to be so helpful."

A trace of a strange smile appeared and disappeared quickly on Antoine's face.

"I am afraid I must hurry. According to the rules of the game one ought to attend to such business immediately. As soon as everything is arranged I shall come back. Will you stay here all day?"

Galois nodded.

In the afternoon Antoine came back to tell Evariste that everything was arranged, since unfortunately he had not succeeded in settling the matter peacefully though he had done all he could. The duel would take place next day at six in the morning. He, Antoine, would come for Evariste at exactly five o'clock. He promised to order a cab and bring two identical pistols. There was nothing, absolutely nothing left for Evariste to do except perhaps practice shooting. The seconds had agreed on a standard type of duel—*à volonté*. One thing more: they also agreed, on their honor, that the whole affair should be kept secret; the names of the adversaries should remain known only to those who would be present next day. They decided not to have a doctor with them, but Antoine's friend, the other second, was a medical student who would be able to help in case of emergency.

Evariste listened in silence, and when Antoine asked what his wishes were, he said that he would like to be left alone until next day at five in the morning.

"It is now four o'clock. Thirteen hours to spend with myself! Thirteen hours in which I am free and can do whatever I like. After that, the hateful ordeal of detested motions, taking pistols, standing firm, facing death bravely, keeping a blank, expressionless face, playing a stupid role before the curtain of life falls down. Then I shall live for a little time in the memory of men. They will remember me! Some with kindness, some with anger. Then a time will come when not only Evariste Galois, but the last thought of Evariste Galois will be dead. Someone, someday, will think about me, and it will be the last time the thought of my name comes to the mind of a living man. I shall live in the memory of my friends even when I am dead. But their impressions will become more and more shadowy; the memory of me will be slowly erased by the sponge of time, until a symbol, a name, some faint trace of a picture remains, and then even that will be forgotten.

"Immortality! It is only through fame that men can fight their destiny of dying and vanishing into oblivion. Only a few are allowed to create, before they die, a new symbol, changing with time to live forever in the memory of men. Does it make any difference for the man who dies whether he is immortal; whether the traces he leaves in life are rich and durable? Does it make any difference for the man who dies whether he is immortal? It does!

"What traces shall I leave of my life? My death is for a petty, stupid cause. I am glad that all of us are bound to secrecy. It is a stupid, miserable death, and the less trace it leaves, the better. But what about my life? What traces will I leave in the memory of men?

"Yes! The people's cause was my cause. I wanted to fight tyranny all my life. I know this sounds bombastic. But we live in a bombastic age. My father taught me to fight tyranny. Father. It's good he didn't know, when he died, that I would follow him in less than three years. Mother? She will cry. She will say that all her life she feared that something like this might happen, that something had suddenly gone wrong with her boy who was placid, gay, obedient, when he was young. Auguste and Alfred. Their grief will be deep and real. Theirs may be the last thought of Evariste Galois.

"Immortality! What right have I to immortality? What did I do for the people whose cause I made my own? I didn't fight in the July days. The fight was won without me and lost without me. What then did I do? I made some speeches; I schemed, planned, I raised the dagger. I wanted to show by a gesture that regicide may be justified. I faced trials; I went to prison. Sainte-Pélagie! I was freed exactly a month ago! What does it all amount to? Nothing. What were the results? None. Did I win the right to immortality in the hearts of the people? I did not. I do not deserve immortality. Yes, true, I am young. Perhaps I could have lived in the hearts of the people if I had been allowed to live longer.

"Immortality! Sometimes it can be bought of life by one moment of heroism. It can be bought by young or old in the right moment when history is willing to sell it. How gladly would I have paid for it with my life. When immortality was for sale, I was a school prisoner. In the time of Louis Philippe, immortality is not for sale.

"There is another immortality. Not that gained by the sacrifice of heart, but by the achievements of the mind. By such achievement I have earned the right to the memory of scientists and mathematicians in France and everywhere over the world where mathematics is taught and understood. But there are not enough traces of my work.

The few papers I have written and that have been printed are short, fragmentary; they do not contain important results. They went unnoticed, and they will be forgotten. The papers I sent to the Academy have never been published. The most important results I carry in my head. And what I carry in my mind will cease to exist when my mind ceases to function and when my heart ceases to beat. The traces of my work can live only if I add to the work of my mind the purely mechanical effort of writing these results with durable ink upon durable paper. Then the traces will remain. Though I have all my results clear in my mind, if I die no one will know that it was I, Evariste Galois, who solved the problems which someone else perhaps will solve again in years to come. Yes, I do want their solution to be connected forever with my name; I want them to be known in mathematics as Galois' theorems, as Galois' methods!

"Immortality! I still have thirteen hours to achieve it. I can still do it. Thirteen hours is much. They are my last thirteen hours. I shall write now the most important results, the outline of my methods, with durable ink on durable paper. To write down everything would require weeks. I have no time. I have no time. I must hurry. No more time left for planning. I must start now—immediately. Yes, I have candles; I have ink and paper; but, above all, I still have my head and in it my brain which will function as long as I am alive. Only I have so little time, I must hurry."

He felt violent hunger. He ran to the janitor's wife and asked her to buy food. What should she buy? Whatever she fancied. It did not make the slightest difference. Then he ran back to his room.

"First I must write letters. Then all the rest of my time I will devote to mathematics. I must hurry. I want to leave a clean memory in the minds of Republicans. They must know that I thought about them and about our cause before I died. Then my scientific testament. All that can be done before five o'clock tomorrow morning. I will leave it to Auguste. He will devote his whole life to making it known. Auguste Chevalier will have Evariste Galois' scientific testament. But this will come later. Now for the letters. There is no time to waste. I must hurry."

He wrote:

May 29, 1832.

Letter to all Republicans:

I beg patriots and my friends to forgive me that in dying I do not die for my country.

I die the victim of an infamous coquette. My life is quenched in a miserable piece of slander.

Oh, why do I have to die for such an unimportant cause; to die for something so contemptible?

I call on heaven to witness that it is under compulsion and force that I have yielded to a provocation which I tried to avert by every means.

I repent having told a baleful truth to men who were so little able to listen to it coolly. Yet I have told the truth. I take with me to the grave a conscience free from lies, free from patriots' blood.

Farewell! It was my wish to give my life for the public good.

Forgiveness to those who kill me. They are of good faith.

E. Galois.

Next, Evariste decided, a personal letter to Lebon and Duchâtelet. He wrote:

My dear Friends,

I have been provoked by two patriots and it is impossible for me to refuse.

I beg your forgiveness for keeping this from you, but my adversaries put me on my honor not to inform any patriot.

Your task is simple: I want you to let it be known that I am fighting against my will after having exhausted all means of reconciliation; and I want you to judge whether I am capable of lying even in unimportant and trivial matters.

Please remember me, since fate did not allow me a life that would make my name worthy to be remembered by my country.

I die your friend,

E. Galois.

Then he wrote a postscript:

Nitens lux, horrenda procella, tenebris aeternis involuta.

The janitor's wife brought packages of food, put them on the chair, the change on the table, made a few remarks about the weather, high prices, the quality of food, and left Evariste's room. Very quickly and hungrily he ate bread with cheese and butter, cold meat, drank two glasses of milk, rubbed his hands on his trousers, and returned to work.

His mathematical manuscripts were mixed with letters and Republican pamphlets. He separated the mathematical notes.

"I ought to read everything, destroy the papers without value, explain others which contain results. But I have so little time. I must leave the papers as they are.

"Now I shall write the letter to Auguste. It will be a long letter. In it I shall formulate all the essential results I have obtained and which I have not yet published; all the results that I am sure are both correct and important. The letter to Auguste will only be a summary and will refer to papers with more detailed theorems and proofs. I may be able to prepare and enclose as many as three papers—two of them on the theory of equations and the third on integral functions. One of the three enclosed papers is ready—the manuscript Poisson rejected. I will have to reread it carefully and make alterations. The second paper on the theory of equations is also partly written, though not in final form."

He remembered how he had planned to print these two papers in a small book.

"If they look carefully through my manuscripts, they will even find a fitting introduction to these two papers." He smiled. "Even if I don't finish everything I intend, there will at least be the letter to Auguste with a summary of everything I want to say. The letter must be written in such a way that years later, if mathematicians should rediscover my results, they will recognize from the letter that it was Evariste Galois who found them first. Yes, I do care about the fate of my name and its immortality. This is my last fight, the fight for immortality; perhaps the only fight I shall win. I shall win my last fight, but I shall never see the sweet fruits of my victory."

He wrote:

May 29th, 1832.

My dear friend,

I have made some new discoveries in analysis.

Then he plunged into mathematics and became technical. He wrote carefully, with the thought that the manuscript might appear in print, that it would be read by mathematicians, discussed, commented on, analyzed. He covered seven pages with words and formulas, and at the end of the mathematical part he wrote:

I want you to know, my dear Auguste, that these topics are not the only ones on which I worked.

Then he briefly mentioned the problems to which he had given much thought of late and explained why he did not report on them in detail:

I have no time and my ideas are not sufficiently developed on that terrain—which is immense.

Then he wrote the concluding sentences:

Please publish this letter in the *Revue encyclopédique*. Often in my life I ventured propositions of which I was not sure. But all that I have written here has been clear in my own mind for a year and it is very much in my interest not to leave myself open to the suspicion that I announce results of which I do not have complete proof.

Make a public request of Jacobi or Gauss to give his opinion not as to the truth, but as to the importance of these theorems.

After that I hope some men may find it profitable to unravel this mess.

Je t'embrasse avec effusion.

E. Galois.

It was still short of midnight when he finished the most important task of his life. Thus he wrote again the date, *"Le 29 mai, 1832."*

He looked at the manuscript which the Academy had rejected. He glanced at the cover and at the names of the referees: Lacroix, Poisson. He felt disgust and at the same time astonishment that disgust should prevail even when he was faced with death.

He decided to read again the eleven long pages. Only half of each page was covered with his writing; the other half formed a wide margin for footnotes, corrections, and supplementary remarks. He wanted to reexamine all the proofs. He felt that his mind was clearer and more penetrating than ever before. If there was anything wrong in his proofs, he would find out today. Today he could throw light upon the problems which had tormented him for months by escaping into darkness. If he only had time!

He read Lemma II, a theorem which had been found in Abel's posthumous paper without proof. Again with disgust he looked at Poisson's note broadly scribbled in pencil. Even in the handwriting he saw something feminine and repulsive. Though he knew the note by heart, he read once more:

> The proof of this Lemma is not sufficient. But it is true according to Lagrange's paper, No. 100, Berlin, 1775.

"I shall fight a small bloodless duel with M. Poisson. Too many duels for one day." He smiled and wrote below the penciled remark:

> This proof is a textual transcription of one given by us in a paper in 1830. We leave here as an historic document—the above note which M. Poisson has conceived it to be his duty to insert.
>
> Let the reader judge.

He read. He read Proposition II and its proof. He saw that the proof was not complete. He saw clearly how it should be changed. He wrote a few lines on the margin but did not like their formulation, crossed them out, and then wrote above: "A few things must be completed in this proof."

A clock struck, announcing that one more hour had passed. He counted. It was twelve o'clock.

He wrote in the margin, "I have no time."

6 : Wednesday, May 30, 1832

Evariste heard a sharp knock at the door. Interrupted in the middle of a sentence, he put aside the pen, piled up all the manuscripts and letters which he had written or prepared during the night, and went to open the door.

Antoine said, "It is exactly five o'clock."

"I am ready."

A minute ago, while he still had his manuscripts before him, he had thought he could go on working forever. Now he felt exhausted. The knock on the door had suddenly drawn all the energy from him, releasing weariness and a desire for sleep.

"If only my adversaries could do me the favor of coming here and putting a bullet through my heart! It would be much easier. Why can't they do that? I need rest, a long sleep. I need it now."

He descended the stairs with wobbling steps, clinging to the banister. Antoine looked at Evariste's pale face, wondering whether its color had been drained away by cowardice or exhaustion.

A cab was waiting. Evariste just succeeded in shaking hands with Antoine's friend who sat in it. He didn't even see the man's face. He threw himself into a corner of the carriage and closed his eyes, afraid that he might faint. The fresh air revived him. He managed to open his eyes and to see Antoine's friend and the tin box that he held on his knees. He thought that the face looked very crude for a medical student and a friend of Antoine's.

Evariste stammered, "You have pistols in the box."

The man opposite him answered, "Yes, would you like to see them?"

Evariste smiled feebly and shook his head.

He was no longer afraid that he would faint. But he wanted to sit quietly, neither to talk nor listen. His companions seemed to understand Evariste's unspoken wish; neither of them said a word.

The carriage passed men on their way to work, dandies on their way to sleep, cheap, untiring prostitutes looking for customers. The horses' hoofs resounded loudly through the half-empty streets.

"Paris is asleep. The smell of horses mixes beautifully with the smell of Paris early on a spring morning. Paris is half dead. But in a few hours Paris will be full of life—as it was yesterday, the day before yesterday, and as it will be tomorrow.

"I am tired. I cannot formulate my thoughts clearly. I'm glad I worked through the night. It is good to know that the manuscripts are piled up in my room. But I didn't do all I intended to do. Of course, it was impossible. But I wrote the letter to Auguste. And there are the two papers on the solvability of equations. The second paper is not quite finished. The paper on integral functions has not been started. Still I did what I could.

"Mathematics! It gave me one last great consolation. It put me into a state in which there is only one desire: for sleep, even if the sleep comes through death. I am not frightened.

"Mathematics! It has given me my only moments of great happiness. Very few are granted such happiness. I must pay for it. This was my true love. I did not think about Eve all night: I lived long, very long; and now I am worn out, tired. My seconds are understanding because they do not chatter to each other and they leave me alone. But I don't like the face of Antoine's friend."

The cab moved through rue Mouffetard toward its southern end where the houses were poorer and less dense, then toward Gentilly through a country road flanked by trees, covered by moving shadows, and cut by brilliant patches.

"The world is showing me all its beauty. She is not cruel, but neither has she pity. She is not sad, but neither does she rejoice because I am dying after only twenty years of life. Twenty years and seven months, to be exact. In five months I would be twenty-one. How much can be done in five months! But even the hour that now divides me from death is a long time. I don't want this hour. I am too tired. I would like to sleep, to add the hour of sleep to the infinity of hours of sleep that are before me. It must be much worse to die when one believes in the immortality of the soul. Then one is doubly afraid because of the prospect of an important examination—much more important than that for admission to the Polytechnical School—and no nonsense allowed. My old behavior would not do. Oh, no, I would not dare to throw a sponge!"

He smiled to himself. Antoine asked the driver to stop. Some fifty yards before them stood another cab.

Antoine said to Evariste, "You stay here and we shall arrange everything."

The seconds left, taking the tin box with them. Evariste closed his eyes, but he was too tired to sleep.

"What will my last thought be? Why am I so sure that I shall die here? There are many other possibilities. Perhaps nothing will happen to me. Then I will have to go through this again and fight the man with the hoarse voice.

"All right! Let's see what comes next. I may be either lightly or heavily wounded. Each possibility has to be considered in conjunction with the possibilities of the second duel. The problem is trivial, uninteresting, and tedious. Not worth pursuing.

"Better to think about the pile of manuscripts on my table. That was intensive work! How often did I start to write a paper that I never finished! I wonder when I would have written out my results, had it not been for this duel? A typically meaningless question. A pity that I did not write the third paper which I mentioned to Auguste. Poor Auguste, he will look for it and won't find it. I ought to have written some remark to spare him worry. He will be distressed thinking that the third paper is lost. Perhaps I shall see Auguste.

"After all, it is not sure that I will be dead in an hour. There are other possibilities. My thoughts are running around circles of different radii, but they all lead to the same tangential point. Such a point can be represented by a particle, and the particle can be represented by a bullet."

Antoine and his friend came back. Antoine said, "Everything is arranged."

Antoine carried a pistol in his right hand. His friend carried the tin box, which he put on the seat of the carriage. Evariste left the cab, and all three of them went into the woods. Some two hundred paces further, they came into a clearing beside a small lake screened from the road by trees. Evariste saw the scenery prepared for their performance. It consisted of two cans screwed into the black earth thirty-five paces apart and two handkerchiefs between them. All four objects lay in a straight line.

The seconds placed Evariste near one of the cans, and he saw Pécheux d'Herbinville near the other; he was faultlessly dressed in black, the collar of his jacket pulled up so that no tie and no shirt were visible.

"He must have a special uniform for these occasions. He must feel very superior when he looks at my brown jacket."

All four seconds assembled now at one spot so that the two adversaries and all seconds formed the three vertices of an equilateral triangle.

Antoine said loudly, "Gentlemen! The choice of places, the choice of pistols was determined by lot. It was also by lot that the privilege was given to me to explain the rules of the duel. According to the agreement between the seconds, a duel *à volonté* will take place between two gentlemen, M. Evariste Galois and M. Pécheux d'Herbinville.

"The distance between the cans is thirty-five paces. The distance between the handkerchiefs is fifteen paces, each handkerchief being ten paces from the can. Each combatant, upon hearing the signal 'go,' is allowed to advance ten paces—that is, from the can to the handkerchief. If the combatants wish, they can advance upon each other, holding their pistols vertical while advancing. The combatant who first reaches the handkerchief must stop and fire. But although one of the parties may thus advance to the limit, his antagonist is not obliged to move, whether he has received the fire of his antagonist or reserved his own.

"The moment one of the combatants has fired, he must halt upon the spot and stand firmly to receive the fire of his adversary, who is not, however, allowed more than one minute to advance and fire, or to fire from the ground he stands on.

"The wounded party is allowed one minute to fire upon his antagonist from the moment he is hit. But if he has fallen to the ground, he will be allowed two minutes to recover.

"M. Pécheux d'Herbinville, are the rules of this duel clear to you?"

"Yes, they are clear."

The black figure bowed in the direction of the seconds and then, more stiffly, in the direction of Galois.

"M. Evariste Galois, are the rules of this duel clear to you?"

Evariste repeated the words and gestures of his adversary.

"You will now receive the pistols from your respective seconds and await my signal."

Two men moved from one point of the triangle to the two other points, handed the pistols to the combatants, and came back to the point from which they had departed.

"Gentlemen, are you ready?"

"Yes."

"Yes."

"Go."

Pécheux d'Herbinville moved calmly toward the handkerchiefs in slow, measured steps, holding his pistol vertically. Evariste stood near the can, motionless, pale, staring into space, spellbound by the approaching black figure. On this black background he saw a fantastically interwoven pattern of shining mathematical symbols.

Suddenly all these symbols vanished, erased by a simple thought:

"I forgot to write to Alfred. This will be a hard blow for the poor boy. I should have written a letter to Alfred. How he looked at me when he came to Sainte-Pélagie. . . ."

Pécheux d'Herbinville reached the handkerchief, leveled his pistol, aimed nonchalantly, and fired. Galois leaned backwards, then recovered, stood erect for a moment, leaned slightly forward, then wavered from right to left like a marionette. Everyone waited tensely to see whether he would succeed in restoring his body to the position of equilibrium, when suddenly he fell stiffly upon his face.

Antoine took out his watch.

"Gentlemen! I shall count two minutes in which the wounded combatant is allowed to return the fire. I ask all of you not to move from your places."

Everyone stood rigidly, staring at the brown figure lying on the earth to see whether it would rise and return the fire. But the figure remained motionless.

"Gentlemen! Two minutes are up. The duel is over."

They all moved toward Galois. Antoine kneeled and tried to turn him slightly. He said to Pécheux d'Herbinville, "He is gravely wounded in the abdomen. I would advise you and your seconds to leave this place immediately. We shall remain here and do our duty."

Pécheux d'Herbinville and his seconds bowed and went toward the road. Antoine's friend laughed aloud when he heard the noise of receding hoofbeats, then with his right foot he kicked Galois' body.

Antoine said sharply, "I do not like your vulgar gestures. Leave him alone."

Antoine's friend said humbly, "Shall we take the pistol away?"

"You are a fool. You must leave everything as it is. We are going now, of course, to look for a doctor. It will not be our fault if it takes us a very long time to find one."

Antoine looked at Evariste and said, "No, he will not last long. I rather liked him."

Then he philosophized:

"The difference between me and you is that you like this kind of work, whereas the most I could say about it is that I don't mind it. Yes, it is true that I enjoy the preparations. They require thought, artistry, brainwork, finesse. But this? It is pure butchery. I don't like blood.

" 'Evariste Galois, a fierce Republican, died in a duel with his Republican friend.' I am sure that M. Gisquet will write something like that in his memoirs. Of course you are not interested in M. Gisquet. You are only interested in your hundred francs. Of course you shall have them. But you will never become an artist. You have the soul of a grocer; you don't know what it means to have the satisfaction of a job well done."

They went in the direction of the road.

Two streams flowed simultaneously into Evariste's consciousness, both growing at the same rate. One was the cruel, hardly bearable pain. The other was the knowledge of where he was and of the events which had brought him there.

He raised his head slightly from the humid earth and called, "Antoine, Antoine!"

The only answer was the joyous sound of birds and the rustling of leaves. He called louder, afraid the birds and leaves would drown his voice:

"Antoine, Antoine!"

No human voice answered. The increasing wave of pain came with the sudden realization that he had been betrayed, that what had happened was quite different from what he had thought. This realization was foggy, like the first idea of a complicated mathematical problem with a far-distant solution. He couldn't even glimpse the solution, because pain covered his field of vision with a heavy, dark curtain through which light could not penetrate.

He cried aloud, bitterly and violently. He cried because he could not stand the pain, because he was lonely; he cried because he was ashamed of men who could do what Antoine and his friend had done. He cried because he felt pity for himself; he cried because the world had shown him, in the last moments of his life, its rotten core.

The cry increased the pain and in turn the pain increased the violence of his cry. The two streams of consciousness began to weaken. Through the noise of his sobs and cries he heard heavy, slow hoofbeats. The earth which he touched brought sounds that became louder and louder. They kept alive the weak and fading spark of consciousness. He waited until the sound was very clear, until it had passed a maximum. He raised the pistol, which was still loaded as it had been when put into his hand. He fired into the air. The sound of the hoofbeats stopped. He called now as loudly as he could, "Help! Help!"

He heard steps nearing and cried again, "Help! Help!"

With the last vanishing spark of consciousness, he saw a rugged peasant's face bending over him. He smiled, clinging to a thought that escaped into darkness: "The world does not show me its rotten core in the last moments of my life."

At noon Evariste awoke in the Hôpital Cochin. His bed was placed in a corner. A screen formed the two other movable walls, isolating his bed from the many patients in the long, narrow hospital room.

Sister Thérèse stood near by and said in a whisper, "You are at the Hôpital Cochin. You are in good hands. I shall go now and call the doctor."

He smiled at her. It was too difficult to speak because of the loud pounding in his ears and the dark spots before his eyes.

A man came who had an egg-shaped bald head and wore glasses. "My name is Paul Sylvester. I am the doctor here."

He took Galois' hand and felt the pulse.

"How did I come here?"

Then he thought, "Why did I ask? It is not important. I can hardly hear my voice. Darkness absorbs words."

"A peasant, a simple man of the people, brought you here. He was very much concerned about you."

Galois smiled to himself. "The doctor is a man of the people. He uses the right language, just the right words. There is more light now. I must say something pleasant to show him that I appreciate what he said and the way he said it."

Words came out with difficulty:

"I am a Republican."

He looked at the doctor's face to see whether it expressed sympathy. But it was difficult to see clearly; something danced before his eyes.

"Answer my questions, but only if you can, only if it is not too much for you."

Evariste nodded.

"What is your name?"

The answer came after a pause. The doctor wrote it down.

"You were in a duel?"

Evariste nodded.

"Would you like to see a priest?"

Galois slowly shook his head.

"We have a priest who is young, understanding, and sympathetic to the Republicans. Perhaps you would like to see him?"

Evariste shook his head.

"Is there someone you would like to see?"

Evariste nodded.

"Today, you don't feel very well." The doctor hesitated. "It is only natural after your experience. Tomorrow you may feel better."

Evariste smiled. He wanted to show the doctor that he understood.

The doctor seemed embarrassed. He said slowly, "If you want to see someone among your friends or from your family, then tell me and I shall send for him. But it ought to be only one person and only for a short visit. Do you know whom you wish to see?"

Evariste nodded.

"Now give me his name and address."

"Alfred, brother, Bourg-la-Reine."

The doctor wrote, "Alfred Galois, Bourg-la-Reine," and said, "Don't worry. I shall do my best, I shall immediately despatch a letter to him by a special messenger. Try to be calm. There is nothing we can do now. Just lie quietly. Sister Thérèse will attend to everything. In a few hours your brother will be here."

He smiled at the doctor and said, "Thank you." Then he added, to show his gratitude by the prolonged effort, "Thank you very much."

Alfred and Sister Thérèse stood at Evariste's bed. She indicated a chair and said, "The doctor allows you only five minutes. You must keep calm, both of you."

Alfred nervously dried his eyes with a handkerchief. Evariste looked serene, smiling at his brother, whose face was full of pain and fear. A sudden stream of tears relaxed his tense face and he cried out, "Who did it to you, Evariste? Who did it?"

Evariste talked very slowly, pausing between sentences, sometimes between words which were hardly audible.

"I can't talk much. I have no time. The King's police. I didn't fire. It is all foggy. The seconds left me. It is all foggy. Who is guilty? Who is not? I don't know. It is too dark to know; *tenebris involuta*."

"Who did it to you? Who did it to you? Tell me and I shall avenge you, I swear."

Evariste shook his head.

"No, Alfred. No revenge."

Words came to his mind which he had heard a long time before, a very long time before. Even the sound of the voice that said them came back clearly now. He tried to repeat these words to Alfred as he had heard them before:

"Don't hate people. It is the system, not the individual. No revenge. No revenge, Alfred.

"I must tell you something. Important. Mathematical manuscripts are in my room. On the table. A letter to Chevalier. It is for you too. Both. Alfred and Auguste. Take care of my work. Make it known. Important."

He felt relieved and listened to Alfred's violent words:

"I swear to you that I'll do it. I shall do everything. I swear to you that they will be published and recognized. I swear to you that if necessary I shall devote my whole life to it."

Now he broke out with a new, more violent torrent of tears. Evariste looked at him with pity and said very slowly, tearing the words out of his body with increasing pain, "Don't cry, Alfred. I need all my courage—to die at twenty."

Alfred tried to calm his sobs by biting the handkerchief and squeezing it into his mouth.

The doctor came in. He stroked Alfred's head, then took his arm and said, "You must go now."

Alfred did not resist. They both went out, and the doctor said to Alfred, "I am your brother's doctor. I wrote you the letter."

Now Alfred cried without restraint.

"You don't know how terrible it is. My brother is a great mathematician, a great man, and a great patriot. You must save him; you must save him. The King's police killed him. The King's police. He told me so. He didn't fire. He told me so with his last breath. You must save him. My brother is a great mathematician."

The doctor did not answer, but gently stroked Alfred's arm.

"Tell me—is there any hope? Is there any hope? There must be. It would be too cruel, too cruel if he should—" He did not have courage enough to finish the sentence. He repeated, "Tell me doctor, is there any hope?"

How many times had he heard the same question, phrased in different ways but always the same! And how many times had he answered the same question with the same words which he repeated now:

"There is always hope while there is life."

"But tell me the truth. Tell me the truth, doctor. Is there much hope?"

The doctor whispered, "Not much."

7 : Thursday, May 31, 1832

At ten o'clock in the morning the doctor came in. He felt Evariste's pulse and said to Sister Thérèse, who stood on the opposite side of the bed, "The pulse is very weak. He is dying."

Evariste felt that someone who had always loved him now held his hand and was speaking to him. A stream of happiness and peace seemed to flow into his body from the hand that touched him. Whose hand was it? Whose voice was it? The voice was melodious, soothing. When he was young, he thought that only angels had such voices. Whose hand was it? Whose voice was it? Why did he ask? Why didn't he recognize the hand and the voice at once? It was so simple, so amazingly simple, so completely obvious. Of course it was his father. How clearly he heard each word!

"Son, dear son, you are tired."

"Yes, Father, I am tired; I am very tired. But I am better now. Your hand helps. Hold my hand and stroke my head. Yes. I feel better now. I feel almost happy."

The doctor gently dropped Evariste's hand.

"He is dead."

Sister Thérèse made the sign of the cross and covered with a sheet the body of Evariste Galois.

8 : June 13, 1909

On July 2, 1832, Galois' friends carried his coffin to a common burial ground that is unknown today. Three thousand Republicans listened to orations that praised Galois' Republican virtues. Seventy-seven years later, French mathematicians, Academicians, officials paid homage to Galois' genius. During these intervening years France fought wars and revolutions, overthrew its kingdom, its second Republic, its second Empire, and the Paris Commune, finally to build and rebuild its third Republic. During these intervening years, Galois' mathematical results were printed, discussed, and taught; they influenced the development of modern mathematics. Time erased many then famous and powerful names. But the memory of Galois grew in the history of mathematics with the passing of years. There it will live forever.

On June 13, 1909, there was a solemn occasion at Bourg-la-Reine. The mayor, the Secretary of the Academy, civil servants, mathematicians, children, citizens, passers-by formed a crowd before a shabby two-story house. A plaque was to be unveiled, stating in simple words that this was the house in which Galois was born. Jules Tannery, professor of the Normal School, read an oration. From windows of neighboring houses, women and children eyed the mildly interesting spectacle. The professor read from a manuscript, but he did it eloquently with vivid gestures, and the audience listened.

"He was born in this house almost a century ago. His father, Gabriel Galois, was one of your predecessors, M. Mayor."

The professor bowed to the mayor and the mayor bowed back to the professor.

"In difficult times, Mayor Galois gave an example of devotion to liberal ideas. He died as a victim of intrigue and calumnies. His wife, née Demante, was a fine and intelligent woman, and she carried a name well known at the law faculty."

He then mentioned Galois' youth at Louis-le-Grand, and his rising passion for mathematics.

"His other passion was a mystic and violent love for the Republic, a Republic perhaps more ideal than his mathematics and too far removed from reality, a Republic for which he was ready to sacrifice his life and, if necessary, those of others. The creations of Victor Hugo are no fictions. Marius and Enjolras are brothers of Evariste Galois."

Then M. Tannery told the story of Galois' life. Yet he did not say that Galois' short life was shaped not by love of a mystic Republic, but by hatred of tyranny—a tyranny as odious as the stench of a prison cell, as perfidious as the treachery of a girl on a police payroll, and as deadly as a well-aimed bullet.

Jules Tannery, in closing his oration, turned to the mayor of Bourg-la-Reine.

"In view of the position that I hold in the Normal School, I have the privilege of saying: Thank you, M. Mayor, for allowing me to make honorable amends to the genius of Galois in the name of the school that he entered with regret, where he was misunderstood, from which he was expelled, and of which he is one of the most shining glories."

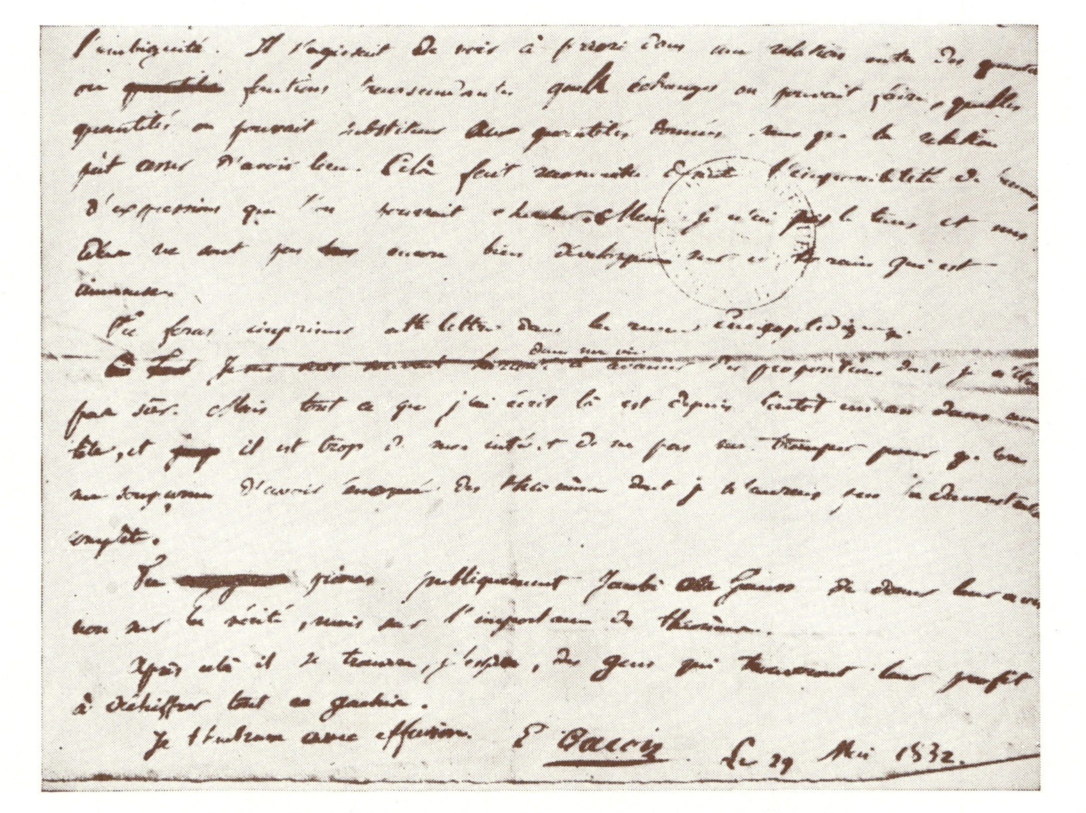

l'ambiguité. Il s'agirait de voir à priori dans une relation entre des quanti[tés] ou fonctions transcendantes quels échanges on pourrait faire, quelles quantités on pourrait substituer aux quantités données sans que la relation pût cesser d'avoir lieu. Cela fait reconnaître de suite l'impossibilité de beaucoup d'expressions que l'on pourrait chercher. Mais je n'ai pas le temps et mes idées ne sont pas encore bien développées sur ce terrain qui est immense.

Tu feras imprimer cette lettre dans la revue encyclopédique.

Je me suis souvent hasardé dans ma vie à avancer des propositions dont je n'étais pas sûr. Mais tout ce que j'ai écrit là est depuis bientôt un an dans ma tête, et il est trop de mon intérêt de ne pas me tromper pour qu'on me soupçonne d'avoir énoncé des théorèmes dont je n'aurais pas la démonstration complète.

Tu prieras publiquement Jacobi ou Gauss de donner leur avis non sur la vérité, mais sur l'importance des théorèmes.

Après cela il se trouvera, j'espère, des gens qui trouveront leur profit à déchiffrer tout ce gâchis.

Je t'embrasse avec effusion. E. Galois Le 29 Mai 1832.

End of Galois' letter to Chevalier written on the eve of his duel.

Evariste Galois' portrait drawn by Alfred Galois from memory after his brother's death.

AFTERWORD

Much truth and some fiction are mixed together in what I have written here. Now I should like to state where truth ends and fiction begins. But this is not an easy task, for I do not know myself. It would be easier to say what is definitely fictional than to say exactly what is true.

In the official registry, there is a birth certificate containing the name of Evariste Galois. Thus one can safely assume that Evariste Galois was born. If one sees letters with the signature of Evariste Galois, if the letters reveal the same handwriting, if, moreover, they reveal an inner consistency in their style and contents, then one can safely assume that these letters were written by Evariste Galois. If one discovers in school records, or in police archives, documents referring to Galois' school or prison life, then one has little reason to doubt that they are genuine. Thus we can safely trust the few known documents that concern Galois' life. But all the documents, all the letters originating during Galois' lifetime and referring to his fate give us only a fragmentary and incomplete picture which must be filled in by the use of less trustworthy sources and by imagination. Wherever I have used sources, my story is as true and trustworthy as the sources on which it is based. Whenever I had to make deductions from known facts, I tried to be as careful and as honest as I could. In the most important matter—Galois' death—my interpretation and conclusions are very different from those of Dupuy, the most celebrated of Galois' biographers. I shall have to treat this matter at some length later. The truth is self-consistent. And in the final appeal, where documents fail, when deduction and imagination must be substituted, this self-consistency remains as the only criterion of truth.

The most important source, quoted and used by everyone who ever wrote about Galois, is Dupuy's seventy-page study. It is a scholarly work, well documented, based on research of the sources, written with warmth and sympathy. But even Dupuy, who never intended to include anything fictional in his short study, had to draw conclusions, had to accept or reject the testimony of some of Galois' relatives, had to go beyond the documented sources which were at his disposal, had to invent, had to formulate his own thoughts, and had to connect events by added links.

This is not surprising. The driest, most scholarly biography must be interpretive too. Otherwise it would be merely a collection of documents. The biographer must deal with the judgments and opinions of contemporaries, with contradicting claims, with evaluations too favorable or too severe, with prejudices and sympathies. No one can describe facts without interpreting them. The mob is bloodthirsty, stupid, and wild for de la Hodde, but noble and brave for Louis Blanc; our interpretations and theories reflect our social attitudes. One can feel apologetic or full of admiration for Galois' revolutionary spirit.

The subjective, personal attitude must be especially prominent when there are as few sources as in the case of Galois. Men who have died at the peak of their fame have had their Boswells. Even if they did not, they still left ample traces of their lives. They usually had wives, mistresses, children, friends, enemies, all of whom jealously preserved letters, scraps of papers, and memoirs referring to the great men. But even then the truth is not easy to ascertain.

To give an example: those who studied and wrote about the life of Victor Hugo can be divided into two groups. One claims that Hugo's wife had a love affair; the other claims that she did not. If the truth is so difficult to ascertain in the case of a man who died old and famous less than a hundred years ago, how much more difficult it is in the case of Galois, who died young and unknown. Usually biographies really start when the hero reaches the age at which Galois' life ended.

As long as he lived, Galois was unknown as a mathematician. He was known only as an ardent Republican. But as such, he conspired and worked in secrecy; he must have done his best not to leave any traces of his revolutionary activities.

Galois' papers, which were found and preserved, were of a mathematical character. All that we know about his revolutionary activities are the traces left in the Parisian daily papers, especially in the *Gazette des Tribunaux,* and in the memoirs of his contemporaries (Raspail, Gisquet, Dumas). It is possible that there were some papers referring to his political activities and that they were destroyed by his family or even by Chevalier.

Indeed Chevalier, in his *Nécrologie,* quotes the following verse, which he says he found among Galois' notes:

L'éternel cyprès m'environne:
Plus pâle que le pâle automne,
Je m'incline vers le tombeau.

(Eternal Cypress shades about me loom
More pallid than autumnal gloom,
My days, I know, approach the tomb.)

I looked in vain for these verses among Galois' papers.

I intend now to discuss each chapter briefly, to say what sources I used and where the story becomes fictional. But, of course, even the part (by far the largest) of the story which is based on sources or documents contains the element of dramatization, which is nearly always inventive.

I. Kings and Mathematicians

The general background of this chapter is based on historical sources.

II. The Rebellion at Louis-le-Grand

A summary of this chapter would present a story of the rebellion at Louis-le-Grand with names, events, dates, and descriptions entirely consistent with the story described by Dupont-Ferrier in his two-volume, scholarly, and heavily documented book. The fictitious parts of this chapter are only the dramatization, Galois' role in the rebellion, and the character of Lavoyer.

III. "I Am a Mathematician"

The formal curriculum as described here is based on documents collected and published by Dupuy. All the remarks of Galois' professors quoted here and later are genuine. Evariste's first contact with mathematics, the impression made on him by Legendre's book, the speed with which he read it, his belief that he had solved the equation of the fifth degree, the beginning of his scientific work, all this is consistent with the story told by Chevalier and in the *Magasin pittoresque*.

Did Cauchy throw away Galois' paper or did he lose it? That Cauchy lost Galois' paper and also that of Abel seems to be improbable.

Galois failed his entrance examination for the Polytechnical School. But the letter quoted here, in which Galois reports this event to his father, is fictitious.

IV. Persecution

The reasons for the suicide of Galois' father and the disturbances at his funeral are described by Dupuy, who learned about them from members of Galois' family. My description is consistent with Dupuy's story. The letter of Galois' father reveals his true reason for suicide, but the letter itself is fictitious.

The information on which the examination scene is based is taken from Bertrand. Did Galois throw a sponge at the examiner's head? Tradition says he did; Bertrand thinks that the tradition is wrong. I have stuck to the tradition, which seemed to me consistent with the pattern of Galois' story and his character.

The discussion between Galois and M. Richard is fictitious. But it accounts for the sudden transition from enthusiasm to coolness in M. Richard's recorded comments. Perhaps it accounts also for the fact that M. Richard did not seem to play any role in Galois' life after he left Louis-le-Grand.

V. In the Year of the Revolution

The story of Galois' expulsion from the Normal School is true, and all documents quoted here and later are genuine. The political and historical background is based on sources. The most important of them is Louis Blanc's book. All the events described here are historical events, but Galois' part in them is fictitious. So is the scene in Pellier's riding school, where weekly meetings of the Society of the Friends of the People actually took place. They were discontinued after September 25, 1831, on which day the National Guard intervened and broke up the meeting.

The scene in which Galois gives the proof of Sturm's theorem is based on information in Bertrand's essay.

VI. "To Louis Philippe"

Up to the banquet scene, the historical background is again based on sources, and Galois' role in the events is partly fictitious. (We know, *e.g.*, that he was, on December 21, 1830, at the Louvre.) The introduction to Galois' paper *On the conditions for solvability of equations by radicals* is authentic, and so is his letter to the Academy in which he urges the referees to say whether they lost the manuscript or whether they intend to publish it. The letter is quoted by Bertrand. The lecture given in Caillot's bookshop is genuine Galois; it is based on one of Galois' notes in his posthumous papers.

Galois enters the historical scene in the banquet at Vendanges de Bourgogne. The banquet and the trial are described in Dumas' *Mémoires* in the *Gazette des Tribunaux,* and in the *Gazette de France.* My description of the trial is almost entirely taken from these sources.

The scene between Gisquet and Lavoyer is, of course, fictitious. About its connection with Galois' story, more later.

VII. Sainte-Pélagie

The chief source for this chapter, and for the description of the *dépôt* and Sainte-Pélagie, was two volumes of Raspail's letters.

Those quoted in this chapter are authentic. However, they are much abbreviated in translation. Their original style is so romantic that some passages would sound ridiculous nowadays if translated literally.

Poisson's judgment of Galois' paper is authentic (quoted by Bertrand) and so is Galois' introduction to his two papers taken from his posthumous notes and published here for the first time. The extract from the *Gazette des Tribunaux* containing the report of the second trial is also genuine.

VIII. Freedom Regained

This chapter must be discussed much more carefully than the others. It contains new deductions which I wish to justify. Let us first collect the facts on which my deductions are based:

1. A bullet entered Galois' cell. This fact, described at length by Raspail, cannot be questioned. One of Raspail's letters also states that all the prisoners were aware that the bullet was not accidental but planned, and that they were indignant when Galois was thrown into the dungeon.

2. The same letter says that Galois was especially badly treated in prison, that he was bullied and pushed around.

3. Galois' prison record shows that he was transferred to the nursing home on March 16, 1832.

4. On May 25, Galois wrote a letter to Chevalier, full of despair and with clear allusions to an unhappy love affair. This letter (quoted in this chapter) was printed by Chevalier in his *Nécrologie.*

5. On May 29, Galois wrote a letter to his two Republican friends, a letter to all Republicans, and his scientific testament. Both the letter to all Republicans and the letter to his two friends were printed by Chevalier in his *Nécrologie.* There the letter to the two Republicans has the following heading: "Letter to N.L. and to V.D."

I believe that it is possible to guess to whom the letter was written. On one of the pages of the manuscript which Poisson had rejected and on which Galois scribbled the famous words, "I have no time," on the eve of his duel, we find the following four names: V. Delaunay, N. Lebon, F. Gervais, A. Chevalier.

One does not need to be a great detective to conclude that these were the names of those to whom Galois intended to write letters on this fateful night. The names N. Lebon and V. Delaunay check with the initials N.L. and V.D. It seems reasonable to assume that F. Gervais was the man to whom Galois wrote the letter destined for all Republicans. Indeed I found the names of Delaunay and Lebon in the *Gazette des Tribunaux* as members of the Society of the Friends of the People involved in Republican trials. F. Gervais is mentioned in Larousse's *Grand dictionnaire universel du XIXe siècle* as an important Republican, a medical man, seven years older than Galois.

(In my description, Galois writes this letter to Duchâtelet and Lebon. This I did so as to avoid the introduction of too many characters about whom we know very little.)

6. Evariste was killed in a duel by Pécheux d'Herbinville.

Dumas, in his *Mémoires,* mentions in one sentence that Galois was killed by Pécheux d'Herbinville, "the delightful young man." There is no other information known and no fact to contradict this. There is no reason to believe that Pécheux d'Herbinville was a police spy. Although Dumas is not very reliable, we must accept his evidence, since it is all we have.

7. Galois was found on the road alone, after the duel, without his seconds. This follows from the notices in the newspapers and from the article in *Magasin pittoresque.*

8. Alfred Galois, Evariste's younger brother, then eighteen years old, saw Evariste in the hospital before his death. Alfred claimed, throughout his life, that Evariste Galois was killed by the King's police. This information, given by Dupuy, seems reliable. Alfred lived through years in which his brother became famous. He must have met many mathematicians when he attempted to draw their attention to Galois' papers, and his views about his brother's death must have become widely known.

This is all we know. Any theory on Galois' death must be restricted by these facts. They form a severe restriction. Let us remember that from the contents of Galois' letters written on the eve of the duel, it follows clearly that Galois was put on his word of honor to keep the affair secret. He knew that he would die, but he did not suspect police provocation. He saw a distasteful love affair as the cause of his death.

It is not easy to invent a theory that will fit all these facts. I do not claim that my story is the only possible solution. But I arrived at it, I should like to say, after three years of familiarity with this problem, during which time I tried to devise a simple but psychologically convincing picture covering all the known facts.

I know that the details are invented and are intentionally vague. But I believe that there is enough circumstantial evidence to prove that the intervention of the secret police sealed Galois' fate. I do not believe that it is possible to fit all the known facts without assuming that Galois was murdered. We know from sources referring to this period that the police knew how to use spies and *provocateurs.* Would they not use their vast machinery to remove, in their opinion, a dangerous, irresponsible, violent, subversive youth who was a threat to the King's life and whom the jury had acquitted? Is it possible to avoid the obvious conclusion that the regime of Louis Philippe was responsible for the early death of one of the greatest scientists who ever lived?

Moreover, there are some further arguments for my theory.

First, we know that M. Gisquet, the Prefect of Police, knew all about Galois' death; that the police were afraid of disturbances; that they prevented the meeting during which a demonstration at Galois' funeral was supposed to be planned. How could they have known all this if police spies were not involved? How did M. Gisquet know that Galois was killed by a friend, as he writes in his *Mémoires*?

Second, I am *not* the first to state in print that Galois was murdered. It is known that after the Revolution in 1848 and during the provisional government, many police spies were unmasked and old plots revealed. It is therefore characteristic that in 1849 a short note on

Evariste Galois published in *Nouvelle annales de mathématiques* begins with the following sentence:

> Galois was murdered on May 31, 1832, in a so-called duel of honor. . . .

Here my circumstantial evidence ends. It is of course possible that some new evidence will throw further light upon Galois' death. But it seems to me very doubtful.

Let us now answer the following question: *What happened after Galois' death?*

Perhaps more interesting than the answer are the sources that provide it. These are two: First, the testimony of M. Gisquet, the Prefect of Police in the times of Casimir Perier, the testimony of one of the men most hated by Republicans. It is given in his *Mémoires,* which appeared in 1840 when no one yet considered Galois a famous mathematician. Second, the testimony of de la Hodde given in his book dealing with the history of secret societies in France at that time. He was a shady figure who pretended to be a Republican until the Revolution of 1848 unmasked him as a police spy. These two sources are identical in their contents.

According to them, a revolution was planned for June, 1832. The Republicans waited only for the proper moment to start it. They thought that the right moment had come when Galois died, and they resolved to make his funeral the occasion for taking up arms.

M. Gisquet begins his report with these notable words:

"M. Galois, a fierce Republican, was killed in a duel by one of his friends."

Does this not imply that the Republicans decided to sacrifice one of them to have a corpse which would inflame the people? But the police, whose hands were clean in this as in all other affairs, as we learn from Messieurs Gisquet and de la Hodde, were well prepared to prevent the outbreak of a revolution, as we again learn from the same source. The meeting to plan the demonstration at Galois' funeral was to be held on June 1, at rue Saint-André-des-Arts at M. Denuand's apartment on which the police placed its seals. But the Republicans broke the seals and proceeded with the meeting. Whereupon the police raided the apartment and arrested thirty Republicans.

But on June 2, all plans for an armed demonstration at Galois' funeral were withdrawn; this we learn again from Gisquet and de la Hodde. Why? On that day General Lamarque died—the hero whom Napoleon on his deathbed had named a marshal of France. Here was a much better occasion to incite the people. Thus the corpse of Lamarque, and not of Galois, stirred the people. Indeed, three days later, on the day of General Lamarque's funeral, Paris was up in arms, barricades were built, and people fought and died for liberty. But Galois was not among those who gave their lives on the barricade St. Merry and whose deeds were described later in immortal words by Hugo. Thus Galois missed the great days of 1832 in which he could have given his life for the people.

Daily papers full of General Lamarque's death mentioned only briefly "the obsequies of M. Evariste Galois, artilleryman of the Parisian National Guard and member of the Society of the Friends of the People," which took place on Saturday, June 2.

Between two and three thousand Republicans were present, among them delegations from different schools. His coffin was carried to the grave peacefully by his comrades, with orations by Plagniol and Charles Pinel, who spoke in the name of the Society of the Friends of the People. Galois' body was put in the common burial ground, and there is no trace of his grave today.

What happened to Galois' scientific works and papers?

Chevalier received these papers from Galois' family. The letter that Galois wrote to Chevalier on the night before his duel was printed in 1832 in the *Revue encyclopédique*. There is no evidence of anyone reading and understanding Galois' scientific testament at that time. We do not know what Chevalier and Alfred Galois did to make Evariste's work printed, read, and known. One of the traces left by their efforts is a copy of a letter written by Alfred to Jacobi; another, Auguste's laborious copies of Galois' papers. It is not known how these manuscripts fell into the hands of Liouville, but it will remain forever to the credit of this famous mathematician that he made an honest and serious attempt to understand Galois' papers and published the most important of them in *Journal de mathématiques pures et appliquées*. We quote from Liouville's introduction:

> The principal objects of Evariste Galois' work are the conditions of solvability of equations by radicals. The author lays the foundations of a general theory which he applies in detail to any equation whose degree is a prime number. At the age of sixteen and on the benches of Louis-le-Grand . . . Galois worked on this difficult subject. He presented successively to the Academy a few papers containing results of his meditations. . . . The referees thought the formulations of the young mathematician obscure . . . and one must admit that this reproach was justified. An exaggerated desire for conciseness caused this defect, which one should strive above all to avoid when dealing with the abstract and mysterious matters of pure algebra. Clarity is indeed most necessary if one intends to lead the reader toward wild territory far from the beaten track. As Descartes said, "When dealing with transcendental questions be transcendentally clear." Too often Galois neglected this precept; and we can understand how famous mathematicians may have thought it proper, by their harsh advice, to turn a beginner, full of genius but inexperienced, back onto the right path. The author they censured was active, ardent; he could profit by their advice.
>
> It is different now. Galois is no more! Let us beware of useless criticism; let us ignore the defects and look for the merits. . . .

We see in these words an attempt to excuse and justify those who never recognized Galois while he was alive. The defense is idle; accusations would be equally idle. The greatness of Galois' tragedy overshadows the problem of the guilt or merit of a few men who read or did not read his papers.

Let us now listen to some gossip about Liouville's publication, told by the mathematician Bertrand in his essay on Galois:

Liouville in publishing the paper which Poisson found obscure, announced a commentary which he has never given. I have heard him say that the proofs are very easy to understand. And when he saw my astonishment, he added, "It is enough to devote onself exclusively to it for one or two months, without thinking of anything else." This explains and justifies the embarrassment stated loyally by Poisson and undoubtedly experienced by Fourier and Cauchy. Galois, before writing up his paper, had reviewed for more than a year the innumerable army of permutations, substitutions and groups. He had to sort and put to work all the divisions, brigades, regiments, battalions and distinguish the simple units. The reader, in order to understand his exposition, must make the acquaintance of this crowd and find his way through it, must learn to see it in the proper light, through long hours of active attention. The nature of the subject demands it. The ideas and the language are new and they can not be learned in a day.

As Liouville wanted to understand the work well, on which he intended to comment, he invited several friends to hear a series of lectures on Galois' theory. Serret was present during these lectures and discussions. The first edition of his *Traité d'algèbre supérieure,* published several years later, said nothing about Galois' discoveries. He said in the preface of his book, that he did not wish to usurp the rights of the master who had instructed him. Fifteen years elapsed before the second edition of Serret's book. Liouville's project of writing a commentary on Galois' work seemed to be abandoned. Serret wrote out Galois' theory for the second edition of his book. He gave to it, I remember, sixty-one pages which were printed and of which I corrected the proofs.

I was astonished that Liouville was not quoted on these pages and when I asked Serret why, he replied, "It is true that I took part in these discussions, but I did not understand a word." However, later, when he saw that it would be difficult to accept this explanation he gave in to the wish of Liouville and suppressed the sixty-one pages. In order to satisfy the typesetter, as the subsequent pages had been already prepared, he wrote an equal number of pages on an entirely different subject.

In 1870, nearly forty years after Galois' death, Camille Jordan wrote a book on the theory of substitutions. He said in the preface, perhaps too modestly, that his book was only a commentary on Galois' papers.

It was this book which drew the attention of the mathematical world to Galois' work. Here are some excerpts from the preface to Jordan's book:

> It was reserved for Galois to place the theory of equations on a definite basis. . . . The problem of resolution, which formerly seemed the only object of the theory of equations, now appears as the first step of a long chain of questions concerning transformation of irrationals and their classifications. Galois, in applying his general methods to this particular problem, found without difficulty the characteristic property of groups of equations solvable by radicals. But in the haste of the formulation he has left without sufficient proofs several fundamental propositions. . . . There are three fundamental notions . . . That of primitivity, indicated already in the works of Gauss and Abel, that of transitivity which appears in Cauchy; finally, the distinction between simple and composite groups. The last notion, the most important of the three, is due to Galois.

At the end of the nineteenth century, the ideas of Galois became generally known among the mathematicians, and their influence constantly increased. In an essay "Galois' Influence upon the Development of Mathematics" written in 1894, the very distinguished and famous mathematician Sophus Lie lists the four greatest mathematicians of the nineteenth century: Gauss, Cauchy, Abel, and Galois. After showing how Galois' ideas penetrated into all branches of mathematics, he finishes with these words:

> After seeing how Galois' ideas proved fruitful in so many branches of analysis, geometry, and even mechanics, one can well hope that they will influence mathematical physics equally. Do not the phenomena of nature present to us only a succession of infinitesimal transformations under which the laws of the universe remain invariant?

In 1906 and 1907 Jules Tannery published most of Galois' remaining posthumous papers. From the scientific point of view these were of little importance as compared with those published before in 1846 by Liouville. In the preface to this edition Tannery writes:

Galois' manuscripts were received by Joseph Liouville from Auguste Chevalier. Liouville left his library and his papers to one of his sons-in-law, M. de Blignières. Mme. de Blignières has occupied herself devotedly with the classification of the innumerable papers of her husband and her illustrious father-in-law. She recovered (not without difficulty) Galois' manuscripts. They were given, together with other important papers, to the French Academy of Sciences.

The lines which follow, some fragments and notes which I publish here, do not add anything to Galois' glory. They are only an homage rendered to his fame, which shines ever brighter and brighter since Liouville's publication.

But Tannery characteristically omitted from his publication a part of one manuscript. We know that when in Sainte-Pélagie, Galois wrote an introduction to two papers on pure analysis, full of complaints, bitterness, accusations, and irony, in which Poisson, the examiners of the Polytechnical School, the men powerful in the world and in science, were attacked and ridiculed. The part quoted here (Chapter VII) in free translation (and slightly abbreviated) is printed now for the first time! It is a strong indictment of a scientific hierarchy which places conceit before humility and arrogance before kindness.

Why did Tannery omit this most characteristic human document? Because, says Tannery, Galois must have been drunk or feverish when he wrote it! The famous mathematician, Tannery, evidently thinks that Galois could not have insulted Poisson and the Academicians without being drunk or feverish. Thus Galois, seventy-four years after his death, is still not allowed to be human, to swear, to be torn by the emotions of scorn and hatred. Seventy-four years after his death he is canonized by the official mathematicians and therefore he must behave like a proper Academician; when he behaves humanly, he must be drunk or feverish.

When Galois died, he was known only as an ardent Republican who loved France, who loved freedom, who fought and hated tyranny. To the mathematician of today, familiar with the words

"Galois group," "Galois field," "Galois theory," he is known as one of the greatest mathematicians of all ages, who died in his youth in a duel. But as long as he lived he was both. His story deserves to be known and remembered not only by mathematicians but by all free men.

BIBLIOGRAPHY

This bibliography does not give a full list of sources and books consulted, but it lists and discusses all those mentioned in this biography and all that contain any new information about Galois. Of the many contemporary English books dealing with the exposition of Galois' theory, it lists only two: the most modern one and the most popular one (as far as I know).

ABEL, N. H., *Oeuvres Complètes. Christiania,* 1839.

ABRANTES, LAURE SAINT-MARTIN JUNOT, Duchesse d', *Memoirs of Napoleon, his court and family.* New York, 1886.

Among the very many books dealing with Napoleon's life, I mention only this one, because it contains the interesting information (quoted in Chapter I) about the influence of Lagrange's death upon the Emperor.

ARTIN, EMIL, *Galois Theory.* Notre Dame, Mathematical lectures, 1942.

This small (70-page) book contains probably the most modern exposition of the Galois theory.

BELL, E. T., *Men of Mathematics.* New York, 1937.

BERTRAND, JOSEPH, *La vie d'Evariste Galois par P. Dupuy.* Printed in *Éloges Académiques,* pp. 329–345, Paris, 1902.

This little-known article contains some information on Galois' life beyond that given by Dupuy. (Galois' letter to the Institute, details of his examination at the Polytechnical School, the story of how the Galois theory became known.)

BIRKHOFF, GARRETT, *Galois and Group Theory.* Osiris, Vol. III, pp. 260–268, 1937.

BLANC, LOUIS, *L'Histoire de dix ans* (1830–1840). Paris 1841–1844. 5 vols.

The most important and exhaustive study of the historical background. Nearly half of this great work covers the years 1830–1832, that is, up to Galois' death.

CHEVALIER, AUGUSTE, *Nécrologie. Revue encyclopédique,* pp. 744–754, Paris, 1832.

This first essay on Galois' life contains the letters that Galois wrote to all Republicans and to his two friends on the eve of his duel.

CRELLE, *Journal für die reine und angewandte Mathematik*. Vol. 1, pp. 65–84, 1826: Vol. 4, pp. 131–156, 1829.

These volumes contain Abel's two papers to which I refer in Chapter IV.

DUMAS, ALEXANDRE, *Mes Mémoires*. Paris, 1863–1865. Vol. 10.

This work contains the description of the banquet at which Galois drank the toast "To Louis Philippe" and also the description of the trial after which Galois was acquitted. This is the only known source that names Galois' antagonist. In one sentence, in passing, Pécheux d'Herbinville is mentioned as the man who killed Galois in a duel.

DUPONT-FERRIER, GUSTAVE, *Du Collège de Clermont au lycée Louis-le-Grand*. Paris, 1921–1922, Vol. 2.

The second volume covers the history of Louis-le-Grand in the period 1800–1920. All my descriptions concerning that school (rebellion in 1824, daily routine, names, M. Laborie's letter) are taken from this work.

DUPUY, PAUL, *La vie d'Evariste Galois. Annales de l'École Normale,* Vol. 13, pp. 197–266, 1896. Reprinted in the *Cahiers de la Quinzaine,* 1903, with an introduction by Jules Tannery.

This 70-page study is the most important source for the story of Evariste Galois. It is a scholarly work containing reprints of many original documents and facsimiles, and reminiscences of those who still remembered Evariste. Yet when the author draws conclusions, he seems to ignore the very facts that he himself collected. Contains also a few factual errors.

EULER, LEONARD, *Éléments d'algèbre*. Paris, 1807.

GALOIS, EVARISTE, *Oeuvres mathématiques,* edited by Joseph Liouville, *Journal de mathématiques pures et appliquées*. Vol. XI, pp. 381–444, 1846.

Includes:

1. *Introduction* by J. Liouville.
2. *Démonstration d'un théorème sur les fractions continues périodiques.* Published originally in the *Annales de mathématiques de M. Gergonne*. Vol. XIX, p. 294, 1828–1829.
3. *Notes sur quelques points d'analyse.* Published originally in the

Annales de mathématiques de M. Gergonne. Vol. XXI, p. 182, 1830–1831.

4. *Analyse d'un mémoire sur la résolution algébrique des équations.* Published originally in the *Bulletin des Sciences mathématiques de M. Férussac.* Vol. XIII, p. 271, 1830.
5. *Note sur la résolution des équations numériques.* Published originally in the *Bulletin des Sciences mathématiques de M. Férussac.* Vol. XIII, p. 413, 1830.
6. *Sur la théorie des nombres.* Published originally in the *Bulletin des Sciences mathématiques de M. Férussac.* Vol. XIII, p. 428, 1830.
7. *Lettre de Galois à M. Auguste Chevalier.* Published originally in the *Revue encyclopédique,* p. 568–576, 1832.
8. *Mémoire sur les conditions de résolubilité des équations par radicaux.*
   This is the manuscript rejected by Poisson.
9. *Des équations primitives qui sont solubles par radicaux.*
   This is an unfinished paper.

In 1897, Galois' papers were reprinted in book form with an introduction by Emil Picard.

Many of the remaining manuscripts were published, with descriptions and comments, by Jules Tannery in the *Bulletin des Sciences mathématiques,* Vol. 30, pp. 226–248, and pp. 255–263, 1906, and pp. 275–308, 1907. They were also reprinted in book form in 1908. The originals of all Galois' known manuscripts are in the Bibliothèque de l'Institut de France. Their photographic reprints are in the possession of Mr. William Marshall Bullitt of Louisville, Kentucky, and in the libraries of Harvard and Princeton Universities. Besides the manuscripts printed by Liouville and Tannery, the collection contains the full introduction to two papers on analysis (reprinted only in part by Tannery), Galois' mathematical notes, some of his school problems assigned by M. Richard, Liouville's notes, and a draft of the letter written by Alfred Galois to Jacobi.

Galois' mathematical notes are often interrupted by drawings: a house, faces, bizarre figures, a chair, many times by an elaborate signature of the author, once by the name, "Gervais," carefully printed, and by the following phrases, *"République indivisible. Unité indivisibilité de la république. Liberté, égalité, fraternité ou la mort. Lyon. Lyon grande ville."*

GAZETTE DE FRANCE, 1831.

The issue of June 17, 1831, contains some details of Galois' trial not printed anywhere else.

GAZETTE DES TRIBUNAUX, 1831.

The issues of June 16, 1831, and of December 4, 1831, contain a detailed description of Galois' two trials.

GISQUET, H. J., *Mémoires de M. Gisquet, ancien préfet de police.* Paris, 1840, Vol. 4.

The second volume (p. 170) contains the story of the alleged conspiracy of the Republicans in connection with Galois' funeral.

HALL, JOHN R., Major, *The Bourbon Restoration.* London, 1909.

HEINE, HEINRICH, *Das Bürgerkönigtum im Jahre 1832.*

A collection of articles. The one dated April 19, 1832, contains the description of cholera in Paris.

HODDE, LUCIAN DE LA, *l'Histoire de sociétés secrètes et du parti républicain de 1830 à 1848,* Paris, 1850.

This book, written by a police spy, contains essentially the same story as Gisquet's *Mémoires.*

HUGO, VICTOR, *The Memoirs of Victor Hugo.* Translated by John W. Harding, New York, 1899.

JORDAN, CAMILLE, *Traité des substitutions et des équations algébrique,* Paris, 1870.

The author states in the introduction that his (667-page) book is only a commentary on Galois' work.

KLEIN, FELIX, *Vorlesungen über die Entwicklung der Mathematik im 19. Jahrhundert.* Berlin, 1926.

KOWALEWSKI, GERHARD, *Grosse Mathematiker.* Berlin, 1938.

LAGRANGE, J. L., *Traité de la résolution des équations numériques.* Paris, 1808. Also: *Oeuvres de Lagrange.* Paris, 1867–1892.

LAPLACE, PIERRE SIMON, Marquis de, *Oeuvres complètes.* Paris, 1878–1892. Vol. 14.

LAROUSSE, PIERRE, *Grand dictionnaire universel du XIXe siècle.*

LEGENDRE, ADRIAN MARIE, *Éléments de géométrie.* Twelfth edition. Paris, 1823.

LIE, SOPHUS, *Influence de Galois sur le développement des mathématiques.*

This study appeared in *La centenaire de l'École Normale 1795–1895,* a large volume on the history of this school. In the same volume, there is also a biographical sketch of M. Guigniault.

LIEBER, LILLIAN R., *Galois and the Theory of Groups.* 1932.

This small book contains a popular exposition of Galois' theory with drawings by Hugh Grey Lieber.

LUCAS-DUBRETON, JEAN, *La Restauration et la monarchie de Juillet.* Paris, 1926.

This very readable book covering the period 1815–1848 mentions Galois, the banquet, and his acquittal.

MAGASIN PITTORESQUE, Vol. 16, pp. 227–228, Paris, 1848.

This volume contains a short, anonymous article on Evariste Galois with a drawing of Evariste made by Alfred Galois from memory, after his brother's death.

MILLINGEN, J. G., *The History of Dueling.* 2 vols. London, 1841.

NOUVELLE ANNALES DE MATHÉMATIQUES, Vol. III, pp. 448–452, Paris, 1849.

This volume contains a short biography of Richard and a note on Galois starting with these characteristic words: *"Galois a été assassiné le 31 mai, 1832, dans une rencontre dite d'honneur, par antiphrase."*

PERREUX, GABRIEL, *Au temps des sociétés secrètes* (1830–1835). Paris, 1931.

This book contains a very extensive bibliography referring to the history of this period and of the secret societies.

PIERPONT, JAMES, *Early History of Galois' Theory of Equations.* Bulletin of the American Mathematical Society, Vol. 4, pp. 332–340, 1898.

PINET, GASTON, *Histoire de l'École Polytechnique.* Paris, 1887.

RASPAIL, F. V., *Lettres sur les prisons de Paris.* Paris, 1839. Vol. 2.

These volumes contain the letters reprinted here in a free and abbreviated translation and also many details about Sainte-Pélagie.

SARTON, GEORGE, *Evariste Galois.* Osiris, Vol. III, pp. 241–259, 1937.

SOURCE BOOK IN MATHEMATICS, edited by David Eugene Smith. New York, 1929.

This book contains an English translation by L. Weisner of Galois' letter to Chevalier written on the eve of his duel.

STENGER, GILBERT, *The Return of Louis XVIII.* Translated from the French by R. Stawell, London, 1909.

TANNERY, JULES, *Discours prononcé à Bourg-la-Reine, Bulletin des Sciences mathématiques,* pp. 158–164, 1909.

This volume contains Tannery's speech made on June 13, 1909, when the plaque on Galois' house was unveiled.

THUREAU-DANGIN, PAUL, *Histoire de la monarchie de Juillet.* Fourth edition, Paris, 1904–1911, Vol. 7.

This work mentions Galois' acquittal after the banquet at the Vendanges de Bourgogne.

VERRIEST, G., *Evariste Galois et la théorie des équations algébriques.* Paris, 1934.

This 58-page pamphlet contains a short sketch of Galois' life and a popular exposition of his theory.